SPATIAL SYSTEMS

To Annemarie, Juliette and Albi

The purpose of models, big or small
Is optimisation, as you may recall
But if you optimise
The model's own size
You might arrive at no model at all

This book is one of a series of Studies in Spatial Analysis stimulated by the theoretical and applied research conducted within the Netherlands Economic Institute, Rotterdam published by Saxon House. Titles in the series are:

J.H.P. Paelinck and P. Nijkamp, *Operational Theory and Method in Regional Economics*
L.H. Klaassen, J.H.P. Paelinck and Sj. Wagenaar, *Spatial Systems*
J.H.P. Paelinck and L.H. Klaassen, *Spatial Econometrics*
Willem Molle, *Regional Economic Development in the European Community*
L.H. Klaassen and N. Vanhove, *Regional Policy, A European Approach*

Spatial Systems

A General Introduction

L.H. KLAASSEN
J.H.P. PAELINCK
S. WAGENAAR

Netherlands Economic Institute, Rotterdam

SAXON HOUSE

Published by

SAXON HOUSE, Teakfield Limited,
Westmead, Farnborough, Hants., England

British Library Cataloguing in Publication Data

Klaassen, Leo Hendrik
Spatial systems.
1. Social policy – Mathematical models
2. Spatial analysis (Statistics)
I. Title II. Paelinck, Jean Henri Paul
III. Wagenaar, S
309.2'01'84 HN18

ISBN 0-566-00263-9

ISBN 0 566 00263 9

Printed by
Itchen Printers Limited, Southampton, England

Contents

Part II Theoretical spatial models

Preface

In this book an aperçu is given of model exercises carried out at the Netherlands Economic Institute in Rotterdam. The main body of its contents was presented as a major discussion paper on a meeting of the World University of Art and Science in Mt. Norikura, Japan in August, 1976.

It expresses the feelings of hope and doubt of the authors of using smaller specific models and large integrated models as tools for describing the spatial aspects of the world we live in today and, more specifically, for using these models as instruments of governmental policy.

The central question raised at the end is whether the growing complexity of society renders larger models useless or rather makes a mathematical approach inevitable. If the reader after reading the book ends up with the same feelings of doubts as the authors, their efforts have not been in vain.

The present volume insists particularly on correct specifications of spatial models. A companion volume (the first on spatial econometrics) develops methodologies for obtaining operational estimates of the parameters of spatial models.

Leo H. Klaassen
Jean H.P. Paelinck
Sjoerd Wagenaar

PART I

GENERAL CONSIDERATIONS AND BASIC CONCEPTS

1. Integral planning; some considerations

Introduction

In recent decades thinking about planning has, in general, clearly shifted from the question whether or not planning is desirable, to the question how planning can contribute in an optimum way to increased well-being. The reason for this shift is caused not so much by evident successes of planning as by the evident shortcomings of so-called 'natural' developments. Increased pollution, social problems, increased traffic and lack of space in many regions, the boom in the demand for education, the evident poverty even in some European regions, all have contributed to the conviction that the no-planning alternative might lead to unbearable consequences for many an individual and certainly would do so for society as a whole. So far, consciousness of these problems cannot be said to have resulted in sophisticated planning systems, however. Social scientists have concentrated strongly on finding operational methods to solve the problems on hand, assisted by developments in computer techniques, but progress has been modest and many an attractive-looking theory has broken down under the rigorous tests it has had to undergo.

And now we find ourselves at a point where the need for effective planning is more urgent than ever before, while the social sciences are not yet able to provide decision makers with the right tools for such planning. In this first chapter some of the causes of this sorry state of affairs will be investigated and an attempt will be made to indicate along what lines research could proceed in order to get us somewhat closer to the objective of integral or comprehensive planning.

The isolated approach

Even at the time when discussions on planning were for the greater part concentrating on the question whether or not planning was desirable at all (a few decades ago, as can be seen in the introduction) there was already a general feeling that the planning of road and rail-road systems, and the provision of social security and special support for the old and handicapped, could not be left to the forces of the free market but would require a systematic approach on government level.

Since then, new fields have gradually been opened up to systematic planning, such as urban development, education and environment. It is realised more and more that in all these fields man can influence the actual developments, although he may not yet have found the right way or, more accurately, the best way of doing so. Characteristically, however, the problems in all these fields are being tackled separately, by isolated bits of planning and government intervention. A few examples may be elucidating.

Educational planning is based upon the demand estimated for different levels of education, given the increase in income and population and the expected changes in the age-structure of the population. From these developments the number of pupils, the number of teachers and professors required for all levels of education and the buildings and equipment for each level can be calculated.

Transportation planning proceeds very much along similar lines. Usually the starting point for a transportation plan is an assumed distribution of population and of economic and social activities over space in a number of future years, e.g. 1980, 1990 and 2000. On the assumption of certain traffic-generation functions the volume of traffic on a given road system can be estimated. As this volume is a function of the quality of the road, it is possible to design a network that serves the demand in an optimum way. For this optimisation procedure computer programmes have been developed. Road networks are then designed according to the results of the calculations and regularly adjusted in subsequent periods.

In regional economic planning, too, many examples of isolated activities can be found. In many a country efforts are made to slow down the growth of investments in urban areas which show signs of overdevelopment, and simultaneously to stimulate the economic growth in backward areas.

To that end, systems of financial support to new firms, or expanding existing firms, have been initiated in underdeveloped regions, in the form of low-interest loans, low land prices or tax exemptions. In the Netherlands, a complementary system was recently introduced, authorising the central government to levy extra taxes from firms newly locating or expanding in the conurbation in the Western part of the country. A similar system may be introduced in Sweden for the Stockholm area. The idea behind such measures is that by giving subsidies and applying the brake on new initiatives elsewhere, one will achieve a new and better – that is, 'more balanced or 'less uneven' – distribution of population and economic activities.

A final example is city planning. In most countries of the Western world urban developments are marked by the influence of the rapid increase in vehicle use. This increase is gravely endangering the position of the city centres, subjecting them to a process of degeneration. Traffic congestion and suburban development seem to lead inevitably to the situation in which many American cities already find themselves, with extensive suburban areas and a central core without population but with many offices, banks, insurance companies, and consequently, a disproportionate need for mass transportation during only a few hours of the day.

To the problems that situation brings, the solution is generally sought in the promotion of public transport, in the hope that it can handle the large number of people who have to be carried twice a day, once from their suburb to the centre and once the other way round. Here, too, the developments are more or less accepted as unavoidable, and the measures taken are no more than attempts to adjust the transport system to what seems to be the natural course of events. Two facts emerge from all these examples.

The first, and perhaps the most important, is that so-called exogenous factors play a prominent part in determining the nature of planning measures. In transportation planning an exogenous set of data is used to lay out the road network. In regional planning it is assumed that investments will continue to be made and that the instruments used will influence only their spatial distribution.

The result is that in city planning, the growth of suburbs and its counterpart, degeneration of city cores, are accepted, and only the transportation system is adjusted, because obviously the private car cannot cope with the traffic situation. In each of the examples quoted there is a marked tendency to adjust to developments that seem exogenous and autonomous. The planning procedures followed are adaptive, the only criteria being that they are to meet in an acceptable way the exogenously determined demands. Also, it must be admitted, that the solutions are quite acceptable, some of them even extremely ingenious, if one accepts the basic assumption of genuinely exogenous developments. But it must also be admitted that this basic assumption itself has rarely been critically examined. It has been accepted as belonging to a different field of interest, and as something accepted for the problem on hand and for its solution.

The development of the demand for education is a datum to educational planners, whose task it is to provide education for those who demand it. Transportation planners have to provide transportation for those who demand it; the same is true of city planners. Regional planners consider investments as understood, and merely try to influence their spatial distribution. Each planner considers the field of his fellow-planners as foreign territories, not falling under his competence. Taking the developments in these 'foreign territories' as understood is the easiest way out of possible conflicts with fellow-planners, so everybody confines himself with painful strictness to his own field. Even minor struggles at the borders are avoided. Each planner is king in his planning domain.

The second fact about conventional isolated planning approaches follows logically from the first. It is that interrelations between different fields of planning are neglected, excluded from study, mistakenly not taken into consideration, or whatever formulation is suitable. The educational planners seldom consider what influence the heavily increased number of university trained people will have on their average income level, if the demand for the skills they have acquired lags behind its supply. Consequently, they neglect the influences that are exerted on demand via the labour market. Transportation planners often forget that the assumed spatial distribution of population and activities that formed the basis for their analysis might not remain in equilibrium with the transportation network they designed on the basis of their analysis. In other words, they have analysed carefully the influence of the spatial distribution of population and economic activities on the transportation network needed to cope with the situation, but have not considered the consequences of the network they have designed for the spatial distribution itself.

The city planners have acted in much the same way. They found themselves in a situation where transportation was the bottleneck for actual developments, and are trying to adjust the infrastructure in such a way that the bottleneck is removed, but they do not wonder much what will be the consequences of such adjustment for city development. Their planning, again, represents mainly adopting the trend without reflection upon the consequences of such an action in the future. They neglect the indirect effect of their behaviour on the state of affairs to which they are trying to adapt.

The integral approach

From the foregoing it is easy to see what are the shortcomings of the conventional approach. To consider every single factor as exogenous that is determined in another field than one's own, means to neglect the indirect effects of one's decisions. Consequently, isolated planning means imperfect planning or at any rate incomplete planning. It may be useful to present again some examples, running parallel to those given in the previous section.

Suppose that a suburban area is developing around a certain city and that the people living in that area are mainly employed in the city centre. Growing car ownership gives rise to a gradually worsening congestion and also to parking space in the city centre becoming manifestly scarce. According to transportation planners, congestion indicates a need for improved infrastructure between suburban area and city core, and the scarcity of parking space indicates the need for a car park in the city centre. Better infrastructure and car parks once constructed will temporarily cope with the demand, but at the same time give a new impulse for suburban growth, and thus eventually create a new demand for better infrastructure and more parking space, and so on and so forth. The transportation planner who just follows demand, is neglecting the effect on suburban growth, and the structural changes of the urban area resulting from such growth. Transportation planning that does not allow for these secondary effects, can produce no more than a short-run plan, because it is not integrated with its neighbour, physical planning.

In educational planning, the promotion of higher education might result in a supply of higher skilled labour considerably exceeding future demand for this kind of labour, a situation that actually exists in Germany, Switzerland, France and Holland. There the stimulation of higher education has resulted in an excess supply of high skilled labour and an undersupply of low skilled labour. The results are well known. While foreign workers are imported in great numbers in order to fill the shortage of low skilled workers, university trained and other high skilled people find it difficult to become employed in a suitable job. Unemployment of highly trained workers goes hand in hand with an excess demand for low skilled workers. The former result provokes a feeling of uneasiness about society as a whole, which does not provide the well-educated with a proper job, while the latter, the excess demand for low skilled workers, causes an influx of foreign workers with all the social problems for these workers themselves as well as for their host country. It is not difficult to understand that the lack of co-ordination between educational planning and labour market planning is to be held responsible for this state of affairs.

A final example is to be found in regional planning. This sort of planning is usually purely economic. Its aim is to increase economic activity in backward areas. This may be industrial activity, it may also be recreational activity. Many regions, particularly in the mediterranean area, try to maximise income from tourism by providing facilities for tourists in the form of apartments, bungalows, campings, caravan parks, etc., without realising that the very volume of their activities may in the long run adversely affect the tourist's desire to come to these areas. We may also point to the conflicting interests of manufacturing industries and tourism; particularly for a country which, like Yugoslavia, is developed mainly near the coast, increasing manufacturing activities might do such harm to the environment, physical as well as visual, that a decrease in tourism will be the result.

Already there are several regions where economic development is playing such havoc with the natural environment that highly skilled workers (with, in general, all who can afford it) are moving out. The resulting shortage of high skilled labour renders the region less attractive for new industries. In fact, even without such shortage, new, modern industries have come to prefer clean, unpolluted surroundings, because their image depends on the quality of the area where their new buildings are to be erected. Obviously, in the areas we were talking about a minute ago, it is the interrelation between economic development and environment that has been neglected, to the detriment of the environment and of economic development itself.

All these examples go to show plainly how closely the aspects of human life are interrelated and why planning for one of them is bound to be unsatisfactory unless the direct and indirect effects on others, and the way the plans boomerang upon that one aspect itself are properly taken into account.

But evidently to take all effects into account from now on, one must have knowledge of their direction and size; in fact, one must know all the cross influences before one can trust oneself to speak of the final effects of a given measure in all relevant fields of human existence. Unfortunately, our understanding, though growing, is far from perfect. Many scholars, aware of the importance of secondary effects and cross influences are striving to gain more insight, and have added many a contribution to our knowledge, but even so it is not adequate for national and integral planning.

Figure 1.1 represents an attempt to picture just the most important relationships that may be assumed between four major areas of interest. The scheme contains, logically, four sub-models. The first sub-model is called SPAMOS (*Spa*tial *mo*del; *s*ocial infrastructure and residential model); it refers to residential location and contains the factors that determine the attractiveness of a given region for residential location, and the factors by which they, in turn, are influenced.

The second sub-model (SPAMOL) represents the labour market as it functions on its own and as it is influenced by factors in the other

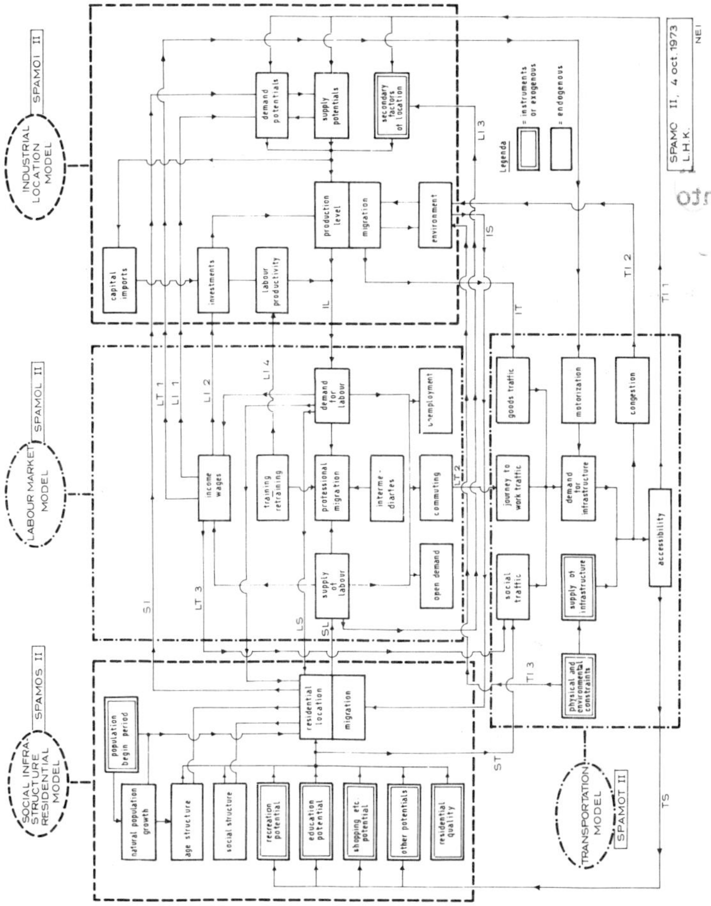

Figure 1.1

sub-models.

The third sub-model is the industrial location model (SPAMOI), the fourth the transportation model (SPAMOT).

The interrelations between these four sub-models as indicated in the scheme will now be briefly described.

1. The social infrastructure and residential model (SPAMOS)

The first sub-model is situated on the left hand side of the graph. The first column represents the factors that determine the attractiveness of a region from the point of view of a resident in i. These factors are taken into consideration by the individual when considering location in i (residential quality, shopping etc. potential, educational potential, recreational potential, social structure and age structure).

The social-infrastructure model is influenced by a number of factors from outside in the following way:

a. the decision to locate or not to locate in i is influenced by the demand for labour in i (LS);

b. the migration between the regions is influenced by the environmental qualities of the regions (IS);

c. the size of the potentials is determined by the accessibilities (TS).

On the other hand the variables of the social-infrastructure model exert an influence on those of other models as

a. residential location (including migration) influences the supply of labour (SL);

b. residential location influences the size of the (industrial) demand potential (SI);

c. the use of the elements of social infrastructure determines the volume of 'social traffic' (ST).

2. The labour market model (SPAMOL)

The essential elements in the labour market model are the processes that lead to balances and imbalances between supply of and demand for labour. The variables, referring to supply and demand, should be given a regional, a social and a 'skill' dimension. Discrepancies between demand for and supply of labour influence the level of incomes and wages. The discrepancies are mitigated by professional migration, which can be promoted by training and retraining programmes as well as by the activities of intermediaries (Labour Offices). The remaining discrepancies result in open demand, unemployment and/or commuting.

The labour-market variables are influenced by

a. social-infrastructure and residential variables, which influences residential location decisions and thus the supply of labour (SL);

b. the production level as a determinant of the demand for labour (IL).

The labour-market variables themselves influence variables in other models

a. the demand for labour influences the decision of residential location (LS);

b. income influences the demand potential (LI 1);

c. income influences the supply of investment goods (LI 2);

d. supply of labour is one of the secondary factors of location (LI 3);

e. income influences the degree of motorisation (LT 1);

f. commuting influences size and structure of the journey-to-work traffic (LT 2);

g. income influences size and structure of social traffic (LT 3).

3 The industrial location model (SPAMOI)

The basis for this model is an interregional intersectoral attraction model in which demand potential and supply potentials influence each other mutually. Secondary factors of location determine, in combination with the demand and supply potentials, the level of production (and thus, indirectly, the migration of firms). Investments, fed by capital imports, influence the production level, a rise of which causes a demand for investments. Investments influence labour productivity, which together with the production level determines the demand for labour. The production level influences the quality of the environments, which itself influences (the structure of) production.

This system is influenced by a number of variables from 'outside':

a. residential location influences the demand potentials (SI);

b. demand potentials are influenced by income (LI 1);

c. income influences the supply of investments (LI 2);

d. the supply of labour influences secondary factors of production (LI 3);

e. accessibilities influence the demand and supply potentials and the secondary factors of location (TI 1);

f. congestion determines the quality of the environment (TI 2);

g. physical and environmental constraints influence the environment (TI 3).

The variables of the industrial-location sub-model affect other variables as follows:

a. the environment influences the migration of persons (IS);

b. the levels of production determine the demand for labour (IL);

c. the industrial location determines size and structure of goods traffic (IT).

4 *The transportation model (SPAMOT)*

Social traffic, goods traffic and journey-to-work traffic together determine size and structure of the demand for infrastructure, which is positively influenced by the degree of motorisation. Physical and environmental constraints limit the supply of infrastructure. Supply of and demand for infrastructure determine accessibilities and congestion.

As can be derived from the foregoing the variables in this model are influenced by

a. the residential location, which influences the demand for infrastructure for 'social' purposes (ST);

b. income, which influences motorisation (LT 1);

c. commuting, which influences size and structure of the journey-to-work traffic (LT 2);

d. income, which stimulates 'social' traffic (LT 3);

e. industrial location, which determines goods traffic (IT).

The variables of the transportation model influence variables of other models in the following way:

a. accessibilities influence the use of the elements of social infrastructure (TS);

b. accessibilities influence supply and demand potentials as well as secondary factors of location (TI 1);

c. congestion influences environment (TI 2);

d. physical and environmental constraints influence the quality of the environment (TI 3).

The mutual influences between the variables of the different models are summarised in figure 1.2. This presentation may suffice to

Mutual influences

of \ on	Social infrastructure	Labour market	Industrial location	Transportation
Social infra-structure		residential location (migration) on supply of labour SL	residential location on demand potential SI	residential location on volume of social traffic ST
Labour market	demand for labour on residential location LS		income on demand potential LI 1	income on motorisation LT 1
			income on supply of investments LI 2	commuting on journey-to-work traffic LT 2
			supply of labour on secondary factors of location LI 3	income on social traffic LT 3
Industrial location	environment on migration of persons IS	production level on demand for labour IL		industrial location on goods traffic IT
Trans-portation	accessibility on use of elements of social infra-structure TS		accessibility on demand and supply potentials and secondary factors of location TI 1	
			congestion on environment TI 2	
			physical and environmental constraints on environment TI 3	

Figure 1.2

demonstrate how complicated even a much simplified scheme of inter-relations is and how much there remains to be done before we shall be able to deal with the problem in all the areas with an acceptable degree of success. Our only hope of reaching that stage, be it sooner or later, lies in concentrating on (1) truly integrated interdisciplinary studies, and (2) the possible consequences that an interdisciplinary approach to the planning procedure as such would entail for the institutional structure of the government executive bodies.

First, however, let us dwell a little longer on the element of space in the interdisciplinary model presented in the SPAMO-scheme.

The element of space

A concept that occupies a central position in the scheme is that of a potential. It is, in fact, important enough in any interdisciplinary model to warrant some extra attention. The idea behind a potential is that an individual who lives in one area is surrounded by other areas, each at a given distance from the point where he lives. Each of those areas offers opportunities for employment, recreational amenities, shopping and educational facilities, etc. Basically, each of the areas and consequently each of the opportunities, amenities and facilities is accessible to him, with the general proviso, however, that the larger the distance is, the smaller the accessibility becomes. For example, the recreational potential of the area where our subject lives, can be defined as the weighted sum of all recreational amenities in his own and all other areas, some distance function serving as a weight. The larger the distance from a given area to his own area, the smaller its accessibility, and thus the smaller the weight attached to the amenities in that area.

In general economics the sacrifices to be made to bridge distance are usually replaced by transportation costs only. Examples for this can be found in the theory of international trade. In spatial economics these sacrifices are treated as a much broader concept. They comprise, except costs, also the time involved needed for bridging the distance, the risks involved in participating in transportation (for goods as well as for persons), the social costs to be made, the resistance in migration sprouting from insufficient information about the area of destination etc. etc. All these elements do play, either all together or in smaller combinations, a very important role in the different sections of spatial economics. We will treat the concepts of distance, potential and accessibility later more extensively. Here it may suffice to state that these concepts form the backbone of spatial economics where 'space' is essentially a multidimensional concept.

The essential function of the potential in models like SPAMO is that it accounts for the relations between regions. When a new recreative amenity is introduced in some region or area, the potential for recreation in all areas will increase. The closer an area is situated to the area where the new facility has been introduced, the more its potential will increase. The same reasoning holds for educational and shopping facilities, employment opportunities, etc.

The distance function determining the weight comprises two important elements. First, the generalised transportation costs, representing the weighted sum of money expenses, effort and time (which, of course, all depend on the means of transport), and second, the influence of these costs on the propensity to make use of a certain amenity. Propensities are interesting, because they appear to differ per motive. The propensity to sacrifice money, time and effort is very low for daily shopping, but high for international flights; low for kindergartens, but high for universities. Obviously, understanding of the propensity to move is essential in physical planning, where precisely the spatial distribution of elements is considered and where consequently the propensities have to be taken into account.

It is interesting to note that distance defined as generalised transportation costs is a function of the quality of both road network and public transport. That means that improvements in the transportation system, too, will increase the value of the potentials in the model.

The concept of the potential is also used in the industrial-location sub-model. There, both the demand for the products of a firm that is located in a given area, and that firm's need of raw materials and semi-finished products, can be expressed as a potential. It should be remarked here that, with firms buying from one another and selling to one another, and with the demand for each firm's goods by all other firms, and the need of each firm for goods of all other firms being expressed as potentials, not only the spatial – or interregional – relations but also the intersectoral relations will thus be accounted for in the model.

Although the model such as it is presented in the SPAMO-scheme might suggest a static structure, it can quite easily become a dynamic model. Into each of the relations a time-lag can be built, should the situation require it. That seems particularly interesting as far as the introduction of infrastructural improvements is concerned. For in that case the ultimate consequences may lag considerably behind the improvement itself, particularly with improvements so extensive that a completely new situation is created. Alpine tunnels might be a case in point.

Three elements, the intersectoral, the intertemporal and the interregional, determine the structure of the model. Add to these its interdisciplinary character, and we have the very completeness that is essential to integrate planning. It is for the social sciences to make this complete model operational. Efforts in this direction by the Netherlands Economic Institute in Rotterdam have at least partially been crowned with success. Considerable progress has been made with the industrial location, labour market and transportation sub-models. The efforts are being continued, in two directions, towards the construction of a compact comprehensive model and towards filling in the large SPAMO-model. It is hoped that the two approaches can eventually be reconciled by a process of disaggregation on the one hand and of aggregation on the other.

In the Netherlands, where physical planning meets with considerable difficulties, there is an urgent need for just such an operational model as described above. In fact, for rational decision-making such an instrument is indispensable.

Institutional planning

From the considerations in this chapter the important conclusion may be drawn that each discipline has a spatial dimension. Developments in one region, whatever the field of study, will always be influenced by and exert influence on what happened in all other regions. Therefore, if we talk about the integration of socio-economic and physical planning, we are not expressing ourselves with adequate precision. By integral planning is meant interdisciplinary planning in which the spatial dimension has been explicitly introduced. In practice, however, physical planning, as belonging to a separate discipline, has always been the privilege of specialised agencies, physical planning bureaux, architectural bureaux and even transportation consultants. On the governmental level as well, there is usually a separate service or institution which occupies itself with urban and other spatial problems. That inheritance from the past forms no mean obstacle to real integrated planning. And yet, the moment our model is divested of its spatial dimension, it breaks down completely, losing its capacity to provide meaningful propositions for spatial arrangements.

The argument presented implies that an attempt should be made to integrate at any rate the higher levels of thinking in physical planning offices. Should even that prove too difficult, then physical planning ought at least to be co-ordinated with other activities that contribute to the construction of the model. But now the difficulties are multiplied, for residential planning is performed in Housing Departments, transportation planning in Public Works Bureaux, educational planning in Education Departments, social planning in Social Departments, economic planning in Economic Departments, etc. and all this not only on the national level but also on the regional and local levels. If we are really serious in our intentions to strive for an integral approach to planning, the present situation must be changed drastically.

Now we do not propose to abolish completely the existing organisation; that would be more irrational than anything else. We do propose that executive staff members of each of the existing organisations, while remaining in office, should form an interdisciplinary steering group which, with the expert support of social scientists skilled in model building, starts to build up gradually an integral model for the area or region under consideration. Thus the social scientist will gradually become familiar with local circumstances while civil servants will step by step gain an insight into the techniques of model building. The same civil servants will see to it that our social scientists remain with both feet on the ground and are not tempted to construct models that are beautiful in theory, but are later proved worthless in practice.

Obviously, the approach we suggest calls for a lot of patience on both sides. Moreover, decision-makers have to remain in close contact with the planning group to make sure that they constantly keep the objectives of local and regional planning in mind. On the other hand the groups should regularly test the objectives presented to them for their consistency and compatibility in the light of the interdisciplinary approach. They must always take into account that an instrument that serves admirably to achieve one goal may quite easily counteract efforts to achieve another, either in the same or in another field. All parties must contribute towards a gradual build up of knowledge. The objective should be to create operational instruments by scientifically sound methods.

At this point it must be pointed out that regional authorities cannot just proceed along an independent course without heeding what happens on lower and higher levels. The integral planning of the region should be consistent with national planning, such as it is being done in the fields of infrastructure, education, etc. It should also take account of the desires and objectives of lower governmental bodies. For planning to proceed smoothly on the regional level, co-ordination both ways, with higher as well as lower authorities, is essential. Besides, there should be co-ordination with regional authorities in neighbouring regions, a requirement that can be the more readily fulfilled as other regional authorities are following the same lines in their approach to integral planning. Regional objectives should be consistent not only from the viewpoint of the region itself but also in an interregional context.

More than any other organisation, regional governments seem indicated to form the backbone of integral planning for a nation. They are close enough to local authorities to guard their interests, and important enough to be taken seriously by the national government. In either sense they have the advantage over one of the other two kinds of governmental bodies.

In the next chapter we will investigate more closely the role of spatial planning in integral planning.

2. The role of spatial planning in integral planning

Introduction

Planning, according to Webster's Third New International Dictionary, means, among other things: arranging the parts of, and devising procedures in accordance with a comprehensive plan for achieving a given objective.

Two things must be ascertained before one starts planning. First, that one is in a position, has the authority, to arrange or to devise procedures; second, that one is clear in one's own mind about what planning, arranging, devising procedures, implies; in other words, that one knows the rules by which to arrange and devise.

A distinction should be made between arranging according to a pre-fixed plan which is supposed to represent the be-all and end-all of all possible arrangements; and trying to find an arrangement that is feasible and acceptable, and which need not be identical with the 'ideal' arrangement.

Direct planning

It seems reasonable to assume in what follows that planning should be understood as 'arranging in a feasible and acceptable way', or as 'devising procedures to accomplish a feasible and acceptable arrangement'. As already said, to undertake such planning one must be in a position to arrange and to devise procedures, and one must understand the rules by which to do it. Two conditions seem interesting to discuss a bit further, because they represent, in fact, the essence of all planning problems.

Who is in a position to do the arranging and devising? That depends greatly upon the manner in which the arranging is done.

In direct positive planning the elements that are objects of planning are directly arranged in a certain order. That is what happens when a room is being furnished: the available furniture is placed about the room in an orderly way. Admittedly the order that is chosen in this case depends on subjective factors, but still it serves as an adequate example of obvious direct arranging: order is created by arranging the elements in a pattern thought desirable.

However, the result achieved here through direct arranging, that is, through the planner's performing himself all the operations needed to create the desired order, can in principle also be accomplished in a negative way. That will be elucidated by another example.

Suppose there are three elements, A, B, and C, to be arranged in an acceptable and feasible way. The space available consists of three cells. Suppose further that the 'ideal order' would be: A in the first cell, B in the second, and C in the third, thus:

A	B	C

The desired result could have been achieved by operating the following 'arranging rules':

1 A, B, and C are to be placed in the three cells 1, 2 and 3;

2 B and C must not go into the first cell;

3 A and B must not go into the third cell;

4 A and C must not go into the second cell.

Such negative arranging by means of marginal conditions would result in exactly the same outlay as positive arranging, though the starting points are entirely different.

The latter, negative, procedure comes much closer to what happens in actual spatial planning than the former. Positive direct planning implies authority on the planner's part to arrange elements directly and positively according to a certain pattern; for the other way of planning no such authority is needed, but the planner must have power to prevent, say, A being placed in the second cell. In the former case the planner either has direct authority to place elements where he wants them, or he is in a position to command somebody else to place them so. In the latter case the desired result is achieved by prohibition rather than by command, which implies that the responsibility for placing the elements lies with someone else.

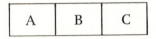

In the practice of spatial planning both forms do occur, but indirect arranging predominates. Direct positive arranging is what a planning government does with elements it controls: roads, public enterprises, universities, etc. Direct negative arranging occurs, e.g. when in an allocation plan a certain site is allotted to industry. Actually, such allocation means no more or less than that it is not forbidden for industries to locate at the site in question. Whether or not an industry is going to locate there the government can control only as far as public enterprises are concerned; in other cases the decision lies with the companies.

Two implications have been made which need further consideration. The first is that the order ultimately achieved does not depend on whether the arranging was done in a positive or a negative way; the second is that prohibitive and/or mandatory stipulations are fully exogenous, or, to express it more correctly, have no bearing on what actually happens.

Imagine that order A-B-C is the ideal one from an aprioristic government point of view, but that from a business-economic point of view B must not be established in cell 2 because it would not be profitable to do so; in that case the ideal arrangement will not materialise because element B will drop out of the planning procedure Now the question arises whether A-B-C can still be looked upon as the 'best' arrangement; in other words whether it remains wise, now that industries fail to show themselves anxious to locate at the planned industrial sites, to plan residential area A and infrastructure C as was provided for under the assumption that B would be realised. If the answer is no, then the whole planning procedure is unsettled. The planned order does not materialise because the prohibitive rules (marginal conditions) have changed the set of elements to be arranged and, consequently, the target arrangement.

Frequently, such a sequence of events leads to a revision of prohibitive stipulations, notably when, in our example, pattern A-O-C seems inferior to a new arrangement B-A-C, in which industrial sites will indeed be used as such. Evidently, the arrangement originally supposed to be the ideal one, can only be so if the elements A, B, and C are viable in cells 1, 2, and 3, respectively. Business must be profitable, the deficits of public transport must not exceed certain limits, a school must have an acceptable number of pupils, etc. In other words, the arrangement one seeks to establish by means of prohibitive or mandatory stipulations must not violate certain boundary conditions. Should these conditions be ignored, then the ideal arrangement, being not feasible, has no practical significance.

Later on we shall refer in more detail to the rules pertaining to any arrangement aimed at; suffice it here to state that, because the arrangement must be realisable, adequate a priori understanding of the criteria an arrangement should satisfy to have practical significance is essential. It seems that so far the arrangements proposed have not always come up to such criteria.

With the statement in the previous paragraph we have answered the question whether prohibitive and mandatory stipulations can be independent of factual developments. Evidently they cannot, for if they have been designed to help establish a certain arrangement, but fail to accomplish that arrangement in practice, they cannot be maintained and will have to be adapted.

In the example previously given it was assumed that certain stipulations had to be adapted because one of the elements figuring in the planned arrangement failed to materialise at the place provided: a negative reason, so to speak, for modifying the stipulations. However, there may be other, positive reasons for considering certain stipulations impracticable. Pressure may be brought to bear upon the planning authority, strong enough to become a positive reason for a change in policy, e.g. the revocation of certain prohibitions.

Suppose that in a certain region an area x^F is taken up by a certain provision, but that there is a tendency of this area to expand to x^{POT}. The planning authority considers expansion desirable only to a limited extent, and decides to set a limit to the growth at $x^M \leqslant x^{POT}$. As soon as the area taken up by the provision reaches x^{POT}, there will arise between x^{POT} and x^M a tension, the strength of which is determinative for the force required to withstand the pressure for further expansion.

The situation then can be expressed by the condition

$$x^M \leqslant \Gamma x^{POT}$$

or

$$\Gamma \geqslant \frac{x^M}{x^{POT}}$$

In this expression Γ is the spatial-planning constraint. If $\Gamma = 0$, there is full prohibition; if $\Gamma = 1$, development is left entirely free. The case of $\Gamma > 1$, which would be one of stimulation, is left out of consideration here.

Now, autonomous growth of x^{POT} results in inversely proportional decrease of Γ, so that increasing resistance will have to be put up against outside pressure. The situation can be relieved by increasing x^M; that will happen when the objections against allotting ever-smaller values to Γ have become greater than those against increasing x^M, that is, against increasing the area considered the maximum admissible for the amenity concerned.

To illustrate what has just been said, two examples may serve. In highway construction the motto was until quite recently, that demand must be followed. Future traffic flows were estimated on the ground of certain assumed developments, and on that basis a road network of a certain capacity was designed (either with or without optimising methods, but that is immaterial here). As road planners used to anticipate growth in building roads, the capacity of finished road sections mostly exceeded the capacity required by traffic at the time. That means that road building was not only demand following, but also demand creating. The over capacity attracted new traffic, which in turn led to the construction of new roads in a later stage. On several occasions the Government has now expressed as its view that such a development must be considered undesirable, especially for certain regions. A rational enough view, or so it seems, but can it be upheld? That depends on the degree to which traffic growth is autonomous. In the foregoing it was assumed that traffic will grow as a result of improved infrastructure. Without denying the truth of that assumption, one may yet point out that traffic also carries an element of autonomous growth, which, even if it cannot make traffic exceed the road network capacity, can and will cause tremendous congestion unless the road capacity is adapted. Cool planners will not let themselves be swayed by such considerations, but will maintain that they precisely want the disease to burn itself out. But whether a Cabinet Minister can afford to be as cool as that does not depend on him alone, but also on the other members of the Cabinet, the Members of Parliament, the electorate, the influence municipalities involved can exert, etc. Moreover, it is not so very clear what the consequences are of not building roads. The models that have been developed to give us an insight into these consequences are hardly the 'pick of the bunch' from our spatial-economic garden, though serious work is being devoted to them. All the pressures and uncertainties, and the doubts arising from them, will probably lead in some cases to a deviation from the original line of conduct. If that happens, the marginal condition is evidently no longer autonomous, but has become dependent on actual developments. Thus, policy becomes an endogenous factor, and the quantity to be controlled by the marginal condition becomes an endogenous variable in the model; this completely ties up with what has been said in the section 'The isolated approach', chapter 1, page 1.

The second example refers to the establishment of shopping stores 'in the meadow'. Quite understandably, public authorities tend to move with extreme caution in this matter. The devastating repercussions on existing city centres and the serious effects on traffic and land use have given rise to a certain common opinion that such establishments may be realised only when no sizeable bad effects as described need be feared. Again, it is questionable to what extent that opinion can be maintained. The advantages – short-term advantages, to be sure – for consumers are considerable. In their eyes easy parking, having everything near at hand, time-saving and often lower prices are essential bonuses which they lose if these shopping centres – which by the way are very rational from a business-economic point of view as well – are not allowed to materialise. Consumers are also apt to reason that the disadvantages will not be all that bad if for instance only one such establishment were to be created near a large town. The environment there is not so vulnerable anyhow; the infrastructure can absorb quite a lot at the hours they want to go shopping, and the influence on the city centre is at most marginal. Since in this case the government's standpoint is not an absolute one either, it does not seem at all impossible that under pressure from potential customers – enhanced, it is to be expected, by experience gained abroad (e.g. in Belgium and Germany) – a number of hypermarkets will be realised anyhow. Where exactly the borderline is going to be drawn, it is difficult to predict. Once the door is ajar, it may be opened a bit further under pressure after all. And then, policy will have become at least partially dependent upon autonomous developments in society.

Indirect planning

If one accepts the line of thought set out in the previous section, one arrives more or less naturally at the question whether indirect spatial planning would not be preferable to direct planning. For if indirect spatial planning means trying, by means of certain instruments, to influence the behaviour of consumers and producers in such a way that they act according to a certain spatial pattern considered desirable, then the developments denoted as 'autonomous' in the previous section will, at least partially, be made endogenous again. If suitable instruments are available – and if they are effective! – spatial planning can be reduced to an indirect process. In such a process elements would not be arranged directly: it would rather be an interaction process with the ultimate objective of reaching a satisfactory arrangement. Knowledge of the mutual influences at work then becomes a condition for a discriminating use of instruments.

However, indirect planning differs from direct planning in more respects than that of the approach alone.

Under the direct approach the planner defines the desired arrangement on more or less a priori grounds, and then seeks to bring that arrangement about. How far it can be realised will only become evident afterwards.

The thinking process implied in the indirect approach, and the knowledge resulting from it, makes it possible to exclude a number of hypothetical arrangements because they are incompatible with social relationships. That is why the latter approach is more efficient than the former provided that, and as far as, the process of thought results in an operationable understanding of those social relationships. That means, however – as will be argued later on – that the objective function must indeed be very general. It has already been pointed out that the idea is to make consumers and producers behave according to a 'desired spatial pattern'; the questions remain who is going to be the one to make out what is 'desired', and how conflicts between producers and consumers are going to be solved. We shall revert to these questions presently, but may briefly mention here that some kind of a compromise will have to be found.

Are there planning rules?

What has been written so far could raise the question if perhaps something could be said a priori about rules to be applied in designing a certain arrangement, in other words, whether there are any criteria by which one spatial plan should be considered superior or inferior to another. However, putting the question like that, we seem to be heading in the wrong direction. For what we really want are certain conditions of well being, in whatever fashion defined. The spatial layout is only one aspect of these conditions, time being the second. Both aspects refer to economics, social structure and environment (or, more accurately, natural capital), three elements that are no separate worlds, but are closely intertwined by many relations. Moreover, when speaking of the spatial aspect, we are referring not only to the spatial arrangement of physical elements – objects of classical spatial planning from of old, as witness expressions like 'physical planning' and 'aménagement du territoire'; in fact, social structure has a spatial dimension too, influenced, to be sure, by the physical spatial layout but by no means identical with it. Spatial physical layout is, in sum, one aspect of societal structure, beside other, non-physical, spatial aspects.

As a general objective for government policy we would propose the development in the course of time of a societal structure that may be expected in the long run to give a maximum contribution towards human and non-human [1] well being. For that objective to be realised, every variable playing a role in the system must have, in a given space, at a given point of time or in a given period, a value that agrees with that societal objective. If that value requires a certain spatial physical layout, that layout must be accomplished; that will be possible because non-feasible layouts have a priori been excluded from the model. Thus, the spatial physical layout is determined implicitly within the system, together with the other structures, a procedure that is followed in principle in modern development plans.

Continuing to reason in this way, one finds that the spatial layout is determined simultaneously with other parts of the system as a function of what could be called 'real instrument variables'. At the same time it becomes clear that direct spatial planning may be exercised only if, and as far as, it does not affect the solution of the system; for instance on a small scale.

From the foregoing it follows that the idea that one spatial physical structure, being preferable to another on purely physical considerations, should for that reason be realised, must be wrong. Physical considerations are included in the optimisation process, and if from that process there emerges a physical layout which, from a purely physical point of view, is inferior to another, that physically inferior layout has been chosen because of the advantages it offers on other scores, advantages which make it preferable from the point of view of societal – higher-order – objectives.

The *Dutch Orientation Note on Spatial Planning* [2] mentions as a general principle of spatial policy 'promoting spatial and ecological conditions such that

a. the essential purposes of individuals and groups in society are realised as much as possible;

b. the diversity, coherence and continuity of the physical environment are guaranteed as much as possible.'

Such a formulation suggests that spatial planning is essential in that it is supposed to have an autonomous influence on societal well being. That would be in conflict, however, with the analysis presented above, which seems to prove that the spatial physical layout is determined along with other aspects of societal structure, and that it is, rather, the reflection of an optimisation process of wider scope.

In the latter conception, specific objectives of spatial policy can hardly be relevant; instead there seem to be only societal objectives, to be pursued with the help of general instruments. Lambooy [3] also touches upon the subject when he says:

> Now all kinds of topics are sneaking into the objectives of spatial planning, topics the discussion of which requires a

much wider framework than that of spatial planning. Underlying societal (or ethical or political) objectives are at present kept in the dark, although it is these objectives that ought to determine what the 'essential purposes' are that are mentioned in the basic objective.

Continuing his argument Lambooy arrives at:

> many objectives that are truly relevant to spatial policy, e.g. sub-objective 3.10, 'aiming at diversity of living environments on a metropolitan scale'; main objective 7.2, 'aiming at limited mobility', sub-objective 7.7, 'aiming at restricted car traffic'.

Although these objectives cannot be denied a spatial aspect, we have some difficulty seeing them as objectives of spatial policy. Limitation of mobility and restriction of car traffic are objectives so fundamental for our society, that they may never be conceived of as the monopoly of spatial planners. If, and to the extent that, the realisation of societal objectives should lead to, say, shorter home-to-work distances, commuting would decrease naturally, but if realisation of societal objectives should call for an enlarged potential labour market area for individual workers, commuting facilities would have to be extended.

Within the objective of spatial policy as defined above, mobility is condemned without previous investigation into the effect of its limitation on all the relevant factors, and what is even worse, without checking whether or not such limitation would be the logical outcome of a general societal optimisation process.

Apparently we keep coming back to the question whether the objectives presented are to be considered objectives of spatial policy or of total societal policy. Consequently, we keep repeating our conclusion that in fact no policy must be allowed to pursue objectives of its own and that all policies should help achieve the objectives that have gained priority in societal evolution. That being so, it will also be clear at once that economic policy cannot have purely economic objectives, nor social policy purely social, and environmental policy purely environmental objectives. For if we look upon society as one system, as an entity in which, in principle, all instruments affect all objectives, then the effects of all instruments will have to be judged simultaneously, and there is no room for independent groups of objectives or instruments in any field whatsoever.

If we look upon spatial planning as an endeavour to influence the physical spatial layout, we must keep in mind, moreover, that this endeavour directly concerns an essential dimension of all quantities included in the system. For spatial planning perhaps even more than for other measures, it is essential to integrate all actions in a larger system and not to let them operate independently.

The question whether or not planning should adhere to certain rules can now also be answered: if the arrangement aimed at on a societal level is superior to the one on the physical planning, spatial planning must operate according to the demands of society and within the range of what is feasible from a societal point of view, irrespective of any advantages of one physical arrangement over another.

We are left with the question what are the ultimate societal objectives. One can, of course, keep these objectives extremely vague. But one wonders: how can these objectives be expressed in a unique variable, to enable policy makers to adapt their policy to situations resulting from spontaneous, 'endogenous' societal processes, as argued above? Perhaps the problem can be solved by considering that when all is said and done, decision makers will have to go by rules that – as indicated before – first and foremost are designed to help solve conflicts. Producers and consumers, for example, make conflicting claims on space (industrial sites versus habitable residential quarters and recreation areas). The ultimate aim, then, is an optimum compromise, which integrates objectives, instruments, preferences relevated by action groups, supposed desires of 'silent majorities', and so on and so forth.

Practical application

In the previous section a description was given of what could be termed, perhaps somewhat euphemistically, 'integral planning'. This attempts to realise societal goals with the tools available, by means of an interdisciplinary model which does full justice to the spatial as well as the time dimension. The model is interdisciplinary because horizontal links between variables from different disciplines receive the same amount of attention as relations between the variables within each separate discipline. Without such a model full integral planning is a fiction.

The aggravating reality is that there is no such model, and even if there were, it is highly questionable if we should ever come to an agreement about its use, or, even more candidly said, if we should ever use it.

In chapter 1 we introduced such a model. It contained four submodels: a residential location model, a labour market model, an industry location model and a transportation model. As said, each of these sub-models falls apart into a considerable number of partial models. Within the residential location model, for example, can be distinguished an education model, a recreation model, a shopping model, a population model, a migration model, etc. In the transportation model there are to be found a modal split model, an infrastructure-optimisation model and an accident model, etc. All these sub-models and partial models are connected with each other to one large interdependent system, in which it is not fixed a priori what are the exogenous, what the endogenous variables. That means that it is not certain in advance either what variables are to be considered as real instruments.

It is an illusion that such a model could be made operational within a short time. It is quite true that models are continuously being developed, and that these models sometimes even have the pretention of being operational integral models. It is equally true that, if ever, computers do a useful job here by proving that the hypotheses behind these models fail to explain the actual developments. It is perhaps for that reason that many people prefer simulation models: among the various kinds of simulation models there are in fact plenty that escape being discarded simply because there is no statistical material available to test them stringently.

Well, if we are honest enough to admit that so far we have not even developed a fully tested partial model, let alone a sub-model, it will become clear that we are still so far removed from a general model that we cannot even begin to think about a policy based on it. Moreover, even if there comes a day when we are in possession of the integrate model, there will probably be fervent discussions about how to weight the objectives, and if and how to operate instruments of unequal political desirability. Suppose that it could be demonstrated with the model that lower taxation of the rich would in the long run contribute more to equal income distribution than higher taxation of the rich; it is not at all sure that the instrument of taxation would indeed be used to that effect, because that could conflict with some short-term target.

Be that as it may, for the time being we are still without the complete model. Much is being done by many in many places, also by physical planners, many of whom are now showing the same enthusiasm for and trust in models as econometrists in the 'fifties, but their efforts cannot be expected to result shortly in an operational societal model.

Now if that last statement is true, the argument presented in the previous pages is for the major part invalid. For how can we blame a spatial planner for having pursued his own objectives, and tell him he ought to work in a wider societal context with general models, if there are no such models? One might just as soon reproach a blind man for not using his eyes. The wider societal context is recognised by all, but the model is lacking.

Looking at the work of spatial planners – who, by the way, recognised the societal point of view much earlier than, for example, economists – in the light of the statement just made, we can largely be reconciled to the way spatial planning is still being done, although new development plans are now being drawn up on a broader base than they used to. And as far as we are not reconciled to that way of doing things, we can only try, in close co-operation between our own and other disciplines, to contrive an improvement by setting up operational models enabling planners to make more rational decisions than were possible so far. In the following chapter we will concentrate on the construction of at least some essential parts of such an integrated model, as introduced in chapter 1.

Notes

[1] Non-human well being in view of the requirements of the natural environment.
[2] *Oriënteringsnota ruimtelijke ordening; achtergronden, uitgangspunten en beleidsvoornemens van de regèring*, Den Haag, 1974, p.99.
[3] J.G. Lambooy, Doelstellingen van het Ruimtelijk Beleid en het Ruimtelijk Systeem in de Oriënteringsnota, *Economisch Statistische Berichten* (Weekly economic paper of the Netherlands Economic Institute in Rotterdam), November 13, 1974, p.1020 ff.

3. A spatial welfare function

Introduction

Integrated analysis of spatial problems requires knowledge about the social welfare function the maximisation of which should be the ultimate goal of government policy. Although in the preceding chapters some scepticism was raised as far as the possible construction of an integral model was concerned, it seems useful at least to indicate in which way an integrated model could be used in designing government policy.

The welfare function used in this chapter is by no means a complete one. It concerns only a part of the integrated model but since it is used only as an example the reader will understand that just because of its simplicity, such a restricted model could, maybe better than a complicated one, show the essence of the approach the authors have in mind.

The analysis will start with a first elaboration of the concept of a potential already introduced in chapter 1. In the next chapters more applications will be presented of this important concept.

Potentials and social welfare function

The concept of a potential used in this section

The concept of a potential or 'accessibility' as it is called in most English literature, is closely connected with the so-called gravity-models, which try to explain a series of spatial interaction phenomenon between two points on the analogy of physical phenomena.

Gravity theory pivots around Newtonian physical law, which says that the attraction exerted by two bodies upon each other is proportional to their masses and inversely proportional to the squared distance between those bodies. In its traditional form the spatial gravity model, based upon the law of Newton, can be written therefore as:

$$t_{ij} = k \frac{w_i^{(1)} w_j^{(2)}}{c_{ij}^n} \tag{3.1}$$

in which t_{ij} is the amount of interaction between two zones i and j, $w_i^{(1)}$ and $w_j^{(2)}$ are measures of the 'masses' belonging to zone i and zone j, and c_{ij} represents the distance, or in more general terms the generalised transportation costs, between i and j; k is a constant of proportionality and n a parameter to be estimated.

A.G. Wilson (1971) shows how – starting from the relation above – a whole family of spatial interaction models can be constructed, by using additional information on the total interaction between area of origin i and area of destination j.

As an example, Wilson mentions a study of the spatial location of purchases in shopping centres, in which the distribution of population and of the purchasing power of the people were given. In that case, the interaction variable t_{ij} is defined as the flow of money from zone i to shops in zone j. The mass-term for zone i is taken to be *the total purchasing power* of inhabitants of i ($e_i p_i = \Sigma_j t_{ij}$), defined as the average purchase per inhabitant of zone i (e_i) multiplied by the population of zone i (p_i). The mass term belonging to zone j is measured as *the size* of the shopping centre in j on the assumption that size gives a good approximation of attractiveness.

The model is:

$$t_{ij} = (e_i p_i) \frac{w_j^{(2)} f(c_{ij})}{\Sigma_j w_j^{(2)} f(c_{ij})} = p_{ij} (e_i p_i) \tag{3.2}$$

in which:

$$p_{ij} = \frac{w_j^{(2)} f(c_{ij})}{\Sigma_j w_j^{(2)} f(c_{ij})} \tag{3.3}$$

For the correct derivation of (3.2) from the traditional model (3.1) we refer to Wilson. Here we are mainly interested in the interpretation of p_{ij}. Wilson interprets p_{ij} as the *probability* of an inhabitant of zone i visiting shops in zone j, but also as the relative utility, obtained by an inhabitant of i by shopping in j.

Elsewhere we adopted the interpretation of probability (Verster and Klaassen, 1973) introducing at the same time the concept of a potential, defined as:

$$p_{ij} = \frac{w_j^{(2)} f(c_{ij})}{\Sigma_j w_j^{(2)} f(c_{ij})} = \frac{\pi_{ij}}{\pi_i} \qquad (3.4)$$

in which π_i represents the potential of zone i for a certain service or facility. In the context of the shopping model mentioned above, π_{ij} is interpreted as *the absolute quality* which an inhabitant of i attaches to the shopping facilities in j. The choice to buy in a certain shopping centre is determined, however, not only by the absolute quality of j, but also by the quality of shopping centres in all other zones, which compete with zone j for the favour of an inhabitant from i. The quality of the shopping facilities in j is weighed against the totality of shopping facilities potentially available to inhabitants of i by taking the ratio between the absolute quality π_{ij} and the shopping potential of zone i ($\pi_i = \Sigma_j \pi_{ij}$). From this it follows that π_{ij}/π_i is the *relative attractiveness or quality* of shopping facilities in j as perceived by inhabitants of i. This factor of relative attractiveness determines the *probability* that an inhabitant of i visits shops in j; or in other words: it determines the percentage of total purchasing power of inhabitants of i spent in shops in zone j.

On the basis of the foregoing we define the (absolute) potential of region i for a certain facility in a general sense as:

$$\pi_i = \Sigma_j x_j \exp\{-\delta c_{ij}\} = \Sigma_j q_j s_j \exp\{-\delta c_{ij}\} \qquad (3.5)$$

in which x_j = size s_j (e.g. area or a similar measure of size) of the facility concerned, weighted with some quality index q_j

$\exp\{-\delta c_{ij}\}$ = a specification of the distance function $f(c_{ij})$, in which c_{ij} represents the generalised transportation costs, determined as the weighted sum of all costs (money, time and effort) of bridging the distance, and δ is a coefficient for the sensitivity to distance, which represents the strength of the influence of the generalised transportation cost on the interaction between i and j.

There are several reasons why the e-function seems an acceptable function to describe the phenomenon of distance decay. Its choice follows in fact logically from the examination of three possible functions, a power function of the shape

$$f(c_{ij}) = a_o c_{ij}^{-a_1} , \qquad (3.5a)$$

a linear function

$$f(c_{ij}) = a_o - a_1 c_{ij} \qquad (3.5b)$$

and the e-function. Together they form the group of simple functions that all imply a negative influence of distance. The power function has as an advantage that it asymptotically approaches zero with increasing distance. However, the function also has a serious drawback, viz. that it approaches infinity for zero distances. This is in practical research not so very important since distances in fact never are zero. Theoretically, however, it is an inelegancy that can be avoided by selecting another function. The linear function has the advantage of reaching a finite value for $c_{ij} = 0$, but here the disadvantage is that the function reaches a point beyond which its value (which can only be positive) is zero. There is no gradual decrease in attraction.

The e-function combines the advantages of both functions. It reaches a finite value for $c_{ij} = 0$ and approaches asymptotically a zero value for increasing distances.

Of course, the extensive use that we will make of the e-function does not exclude the possibility that for certain phenomena a more complicated function is more suitable. In commuting it is not unlikely that people prefer to live at a certain distance from their work over a situation in which their work is directly next door. In this case a Tanner-function of the shape

$$f(c_{ij}) = c_{ij}^{a} \exp\{-\delta c_{ij}\} \qquad (3.5c)$$

could be used. This function is first increasing, reaches a maximum for $x = \frac{a}{\delta}$ and then decreases continuously. It is obviously a more general form that degenerates to the more simple e-function for $a = 0$. In this section we will use the last mentioned simple e-function

It appears then, that a potential is the sum of the value of a specific variable in all regions considered, weighted with distance and quality.

It is interpreted as the degree to which region i is provided with a certain service or facility within region i itself as well as in all other sub-regions of the whole research-area. In other words: π_i represents the degree to which the facility concerned is within the reach of the inhabitants of region i (hence the concept of accessibility).

Social welfare function

Welfare is not measurable in the common sense of the word, i.e. on a proportional scale with a common unit yardstick. In general the concept of welfare represents the degree to which an individual or a group of individuals is of the opinion that his or their needs are satisfied by the available means. Needs as well as means include material and immaterial, measurable and 'non-measurable' items. However, though welfare as such is not measurable, it is possible to say whether or not an individual or a group of individuals prefers situation A to situation B. In such a case it is commonly said that welfare in A is greater than welfare in B, but the correct meaning is that the individual or the group, if given the choice, would choose A rather than B from which one refers that he ranks A higher than B: (see de V. Graaff, 1967). The preference ordering of the population as a whole is described in a *social welfare function* or collective preference function, a concept analogous to the utility function in the theory of consumer behaviour. Such a function gives a definition of welfare. The arguments of the welfare function are all the factors which can be considered determinants of welfare. Naturally, there are not only economic determinants, but also social, cultural and psychological ones. Not all of them are quantifiable by the same method, some are not or only in an indirect way quantifiable. We find ourselves compelled to choose the most important determinants, in the hope that with their help we shall be able to develop a sort of functional definition of welfare. Of course, such a definition cannot escape being somewhat biased.

For choosing the determinants reference can be made to studies about migration and migration motivations (see: Schouten, 1975-a and Ter Heide, 1965). Every decision to migrate is based upon the evaluation of interregional differences in the quality of life, differences which Ter Heide discerns in four spheres of life:

1 the economic sphere or sphere of work;
2 the sphere of housing (relates to the house itself);
3 the sphere of facilities (the geographical environment of the house);

4 the sphere of social relations (the social environment of the house).

In the following we will make use of three spheres, viz. the spheres of housing and facilities, and the economic sphere. The more recent term 'livability' has more or less the same meaning as Ter Heide's housing factors. As facilities are counted medical care, education, culture, recreation, social servicing, etc.

These variables will be formulated as potentials, in the way described before, for we are interested not only in the physical presence of a facility and in its size (and, perhaps, its quality) but also in the place where that facility is located and in the sacrifices (in money, time and effort) required for using them. Using a potential, one can do justice to these spatial aspects. That idea will now be elaborated somewhat further.

Let us assume that it is possible to construct potentials for all factors relevant to the social welfare function and for each region or urban area distinguished. Let there be v = 1, . . . , V facilities and i = 1, . . . , I = J regions; then the potential of facility v in region i is defined as:

$$\pi_i^v = \Sigma_j x_j^v \exp\{-\delta^v c_{ij}^v\} \qquad (3.6)$$

in which the meaning of the symbols is the same as before. [2]

Next, the potentials found are used to construct a social welfare function. Let us assume, as before, that welfare is determined by three potentials, for instance, a housing potential, a facilities potential and an employment potential. The social welfare function of region i can then be represented by

$$w_i = w(\pi_i^h, \pi_i^v, \pi_i^w) \text{ for } i = 1, \ldots, I \qquad (3.7)$$

which implies that the welfare of region i is a function of all the distinguished components of economic and socio-cultural infrastructure.

The specification of the function is assumed to be such that different combinations of housing, working and facilities can lead to the same welfare position (principle of substitution), and also that the continued rise of one of the potentials, ceteris paribus, will cause a decreasing rise in welfare (declining marginal rate of substitution or decreasing marginal 'returns') [3] As a frame for thinking we tentatively take a Cobb-Douglas-like function:

$$w_i = (\pi_i^h)^{a^h} (\pi_i^v)^{a^v} (\pi_i^w)^{a^w} \qquad (3.8)$$

The form of this function satisfies the conditions required, if

$$\frac{\partial w_i}{\partial \pi_i} > 0 \text{ and } \frac{\partial^2 w_i}{(\partial \pi_i)^2} < 0 \text{ which is the case if } 0 < a^h, a^v, a^w < 1 \qquad (3.9)$$

But equation (3.8) also shows some typical characteristics. Under the potentials approach, the welfare in region i appears to depend not only on the availability of facilities in region i itself, but to be equally influenced by facilities in all neighbouring regions as far as they are accessible. In the next section this interregional interdependence will be treated further.

A second element which complicates equation (3.8) is the possible interdependence of the potentials that are arguments in the welfare function. It will be clear that some potentials are not independent from each other; they also may be functions of the same variable. When that is true, a change in a potential π_i^v, as a consequence of a conscious governmental policy, will influence all the other potentials that are functions of π_i^v. Formally the foregoing can be represented by:

Take: $$w_i = (\pi_{1i})^{a_1} (\pi_{2i})^{a_2} \qquad (3.10)$$

then: $$\frac{\partial \pi_{2i}}{\partial \pi_{1i}} \gtreqless 0 \qquad (3.11)$$

If equation (3.11) represents a *positive* relationship, then both the direct and indirect effects of governmental policy (resp. $d\pi_{1i}$ and $d\pi_{2i}$) are positive, so that the positive change of welfare becomes extra great. These effects are in addition passed on to surrounding regions owing to the interregional interdependence. Such a situation of positive influence could occur, e.g. if the regional stimulation policy of the government, meant primarily to create employment, via the construction of social infrastructure, should increase the recreational potential as well.

In a *negative* relationship the welfare benefit to be gained from a rising potential will be reduced by the decline of one or more other potentials. That could happen if extension of employment were damaging to the environment. In short, it follows that an understanding of the relatedness of potentials and of interregional interdependence is of great importance in determining the welfare effects of government measures. Before we continue this subject, a few more short comments on function (3.8) will be given.

The a's in this function represent the weights of each of the determinants of regional welfare. It will be noticed that equation (3.8) assumes that the same a's apply to all regions and to all groups of population. Seemingly we assume that regions are homogeneous in the demographic and social structure of their population. Generally, however, preferences for the various welfare determinants depend on age group and social group, so that for each relevant group a separate welfare function must be defined. If, on the other hand, we may assume that people belonging to the same group will react essentially in the same way, irrespective of the region where they live, then we may also assume that for such a group the a's will be the same in each region (or: will not be region-specific). That implies that the demographic and social structure of the population in a region are seen as important variables for explaining the regional differences in preference-orderings and the changes therein over time. At a later stage different welfare functions will have to be defined per group in each region. Here this aspect will not be treated in detail.

Little attention was paid so far to possible differences in the preferences of *the population* on the one hand and *the government* on the other hand. But there are arguments to distinguish between social welfare functions representing the preferences of the population and such functions representing the priorities of the government. For a discussion of this point we refer to an earlier paper, (see van den Berg, L., Klaassen, L.H. and Vijverberg, C.H.T., 1975).

A difficult problem with welfare functions is the determination of the a-coefficients. Because it is impossible to make a direct measurement of w_i, it is also impossible to make a direct estimate of the a's with, for example, multiple regression analysis. Elsewhere it was suggested that the problem can be solved with either an *explicit* method using results of public opinion polls of relative preferences (see Schouten, C.W., 1975-b) or an *implicit* method making estimations of the a's on the basis of ex post 'revealed preferences'(see Klaassen and Verster, 1973; Nijkamp, 1970 and 1975; Paelinck, 1973; Hordijk, Mastenbroek and Paelinck, 1974, as well as literature quoted there). In the next subsection we shall also give a short description of an implicit method in which the preferences of the population are seen as expressed in *internal migration movements.*

We shall assume that the distribution of migrants among possible areas of destination corresponds to the relative differences in regional welfare. Stated otherwise: we shall assume that the internal migration flows reflect the interregional differences in welfare. Every migration decision is preceded by an appreciation of the relative welfare level of the possible area of destination, account being taken of the distance barrier due to financial and socio-psychological factors.

The foregoing remarks mean that we postulate a welfare function for the population which has the same form as the welfare function in equation (3.8).

$$w_i = (\pi_i^h)^{\lambda^h} (\pi_i^v)^{\lambda^h} (\pi_i^w)^{\lambda^w} \qquad (3.12)$$

Taking this equation as basis, we can define the relative attractiveness of area j to an inhabitant of area i as:

$$\frac{a_{ij}}{a_i} = \frac{(\pi_j^h)^{\lambda^h} (\pi_j^v)^{\lambda^v} (\pi_j^w)^{\lambda^w} \exp\left\{-\beta d_{ij}\right\}}{\Sigma_j (\pi_j^h)^{\lambda^h} (\pi_j^v)^{\lambda^v} (\pi_j^w)^{\lambda^w} \exp\left\{-\beta d_{ij}\right\}} = \frac{w_j \exp\left\{-\beta d_{ij}\right\}}{\Sigma_j w_j \exp\left\{-\beta d_{ij}\right\}} \qquad (3.13)$$

where a_{ij} = the attraction of j for the average inhabitant of i

a_i = the sum of the attraction of all areas for an inhabitant of i

d_{ij} = the total generalised (financial and social) costs of a move from i to j

$\lambda^h, \lambda^v, \lambda^w$ = relative weights of the attraction potentials

β = distance sensitivity.

On the ground of our hypothesis of migration behaviour, viz., that the flow of migrants from i to j is a function of the relative attraction of region j, weighted with the generalised migration costs, we can write (see also Klaassen and Verster, 1973):

$$m_{ij} = \frac{a_{ij}}{a_i} m_{i*} = p_{ij} m_{i*} \qquad (3.14)$$

where m_{ij} = the flow of migrants from i to j

$p_{ij} = \frac{a_{ij}}{a_i}$ = the probability that an inhabitant of i will migrate from i to j

m_{i*} = the total flow of migrants from i to all other regions ($m_{i*} = \Sigma_j m_{ij}$).

We note that equation (3.14) holds for all i and j, therefore also for m_{ii}. This last symbol represents the number of people who do not migrate but go on living in i. Their number has also been incorporated in m_{i*}, which therefore equals the number of people who stay in i, plus those who move either from i to j or somewhere else.

By definition this number equals the total population of i, which implies that (3.14) can also be written as:

$$m_{ij} = p_{ij} P_i = \frac{(\pi_j^h)^{\lambda^h} (\pi_j^v)^{\lambda^v} (\pi_j^w)^{\lambda^w} \exp\left\{-a d_{ij}\right\}}{\Sigma_j (\pi_j^h)^{\lambda^h} (\pi_j^v)^{\lambda^v} (\pi_j^w)^{\lambda^w} \exp\left\{-\beta d_{ij}\right\}} P_i$$

$$= \frac{w_j \exp\left\{-\beta d_{ij}\right\}}{\Sigma_j w_j \exp\left\{-\beta d_{ij}\right\}} P_i \qquad (3.15)$$

In words, equation (3.15) expresses that the flow of migrants from i to j is determined by the relative weighted attraction potentials (or the relative regional welfare) of j with regard to i, and is proportional to the total population of i.

If we succeed in quantifying all potentials and distance functions, it will become possible to estimate by some statistical regression method the λ-coefficients, which are the weights attached by the population to the determinants of its welfare. With the help of the λ's we can identify the marginal preferences of the population and make other use of the social welfare function in which they appear. [4]

Government expenditure, potentials and welfare

Governmental welfare policy

The concepts of the potential and the social welfare function developed earlier can be used to obtain a better insight into the regional or urban welfare policy of the government. It is the central government's responsibility to promote welfare in the whole country by means of the instruments of regional policy at its disposal. Because the welfare of a country's population is given concrete form in smaller geographical units, the government, in its policy towards the various regions of the country, will try to make the regional welfare differences as small as possible, or at any rate to ensure to every region a certain minimum level. One of the instruments available to the government is the amount of money spent on the components of economic and socio-cultural infrastructure, in other words, the public expenditure directed towards the increase of the various potentials that are determinants of regional welfare.

The foregoing does not imply that the government should realise in all regions exactly the same combinations of components of the regional welfare function. It does mean that the government should try to influence the level of relevant potentials in such a way that the resulting combination brings the social welfare in every region up to a certain desired standard. That statement implies that all elements of the welfare function can in principle be substituted and that different combinations can represent one and the same welfare level. The specification of function (3.8) corresponds with this implication.

If the government has a certain annual budget at its disposal for increasing potentials, it is confronted with a two-sided selection problem:

a. to decide upon the spatial allocation of governmental funds to the various regions, giving priority to regions with a relatively low welfare;

b. to decide upon the functional allocation of funds within each region to the various welfare components, giving preference to components with the highest contribution to welfare.

The regional as well as the functional allocation of public funds is determined according to the priorities of the government, embodied in its social welfare function. In this governmental selection function which relates to the welfare of all regions together, the total welfare of all regions will be weighed as well as the various welfare components within each region. Both sorts of weighing are directly related to the actual regional situation. The increase of welfare in a lagging region contributes most to national welfare, and the rise of a relatively low potential within a region has extra priority.

In the last section it was already expressed that the selection process is complicated by different forms of interdependences between potentials and between regions. In the next subsection attention will be paid to the interregional interdependences and the relations between potentials and governmental expenditure will be dealt with.

Potentials and government expenditure

Government expenditure is taken here as an instrument to increase potentials. The various ways in which a potential can be increased by means of public expenditure are evident from the potential formula:

$$\pi_i = \Sigma_j q_j s_j \exp\left\{-\delta c_{ij}\right\} = \Sigma_j x_j \exp\left\{-\delta c_{ij}\right\} \qquad (3.16)$$

which indicates that public outlays for a certain facility can be related to

a. the quality of the facility, the size remaining the same;

b. the physical size of the facility at equal quality;

c. the quality of the transport infrastructure. [5]

Government expenditure is distinguished into current expenditure and capital expenditure or investments. In the following we shall restrict ourselves to public investments for net expansions of the physical size of facilities at equal quality. In that case the increase of a potential equals:

$$\Delta\pi_i^V = \Sigma_j \Delta x_j^V \exp\left\{-\delta c_{ij}\right\} \qquad (3.17)$$

Put:

$$\Delta x_j^V = k_j^V i_j^V \qquad (3.18)$$

in which k_j^V represents the additional size of facility v that can be produced in region j with a unit-of-investment amount i_j ; then it follows that:

$$i_j^V = \frac{1}{k_j^V} \Delta x_j^V = p_j^V \Delta x_j^V \qquad (3.19)$$

in which p_j^V represents the price of a unit added to the size of facility v in region j in terms of the governmental capital outlay needed.

From equation (3.2) it also appears that the relative potential of region i is determined not only by public expenditure in that area but also by the corresponding public expenditure in all other (relevant) regions, since π_i^v, which appear in the denominators (see equation (3.16)), is the sum of all π_{ij}^v's. Such a spill-over or interregional interdependence is essential for a potentials-approach. When, for instance, region i invests in educational facilities, not only will the educational potential in i, and thus the welfare level of i, increase, but the educational potentials and thus the welfare level of all other regions will also be influenced by those investments, though only to the extent to which the educational facilities of region i are accessible to other regions.

Reversely, the welfare of i is influenced by investments in, for example, recreational facilities in another region, provided that the costs of transport from i to that region are not prohibitive. The picture would be very complicated, if attention should also be paid to a mutual dependence of potentials. For the sake of clarity we shall abstract from this sort of interdependence.

In principle the government can spend the total net amount of investments on all potentials in all regions, so that, using the same terminology as in equation (3.7), we may write for the government's budget relation:

$$\bar{y} = \Sigma_j \, (i_i^h + i_j^v + i_j^w) \qquad (3.20)$$

or :

$$\bar{y} = \Sigma_j \; p_j^h \, \Delta \, x_j^h + \Sigma_j \; p_j^v \; \Delta \; x_j^v + \Sigma_j \; p_j^w \, \Delta \; x_j^w \qquad (3.21)$$

Simplified, it can be stated that unit investment prices in all regions are identical per facility, so that (3.20) changes into:

$$\bar{y} = p^h \; \Sigma_j \; \Delta \; x_j^h + p^v \; \Sigma_j \; \Delta \; x_j^v + p^w \; \Sigma_j \; \Delta \, x_j^w \qquad (3.22)$$

We will investigate the problems of governmental welfare policy somewhat closer in the next section.

Optimum allocation of welfare investments

Approach for two facilities and n regions

Assume that the welfare function for the i-th region consists of two elements π_{1i} and π_{2i}, defined (as before) as

$$\pi_{1_i} = \Sigma_j \, x_{1_j} \exp\left\{-\delta_1 c_{ij}\right\} \tag{3.23}$$

and

$$\pi_{2_i} = \Sigma_j \, x_{2_j} \exp\left\{-\delta_2 c_{ij}\right\} \tag{3.24}$$

The values for π_{1_i} and π_{2_i} at the beginning of the year under consideration are indicated by $\pi_{1_i}^{(o)}$ and $\pi_{2_i}^{(o)}$, respectively.

In the analysis it is assumed that the attraction element in each region j is defined by x_j, regardless of the number of people using the facility. Obviously this assumption restricts the validity of the analysis but enables us to keep the argument relatively simple. A more complete analysis could be made by assuming that the attractiveness of x_j declines (or increases) with the number of people attracted by it

We now assume that the government divides a given sum of money among the regions and the (two) facilities in such a way that the increase in the welfare of all regions together is maximised.

Write:

$$w_i = (\pi_{1_i})^{a_1} \, (\pi_{2_i})^{a_2} \tag{3.25}$$

and consequently

$$\phi = \Delta \, w_i = (\pi_{1_i}^{(o)} + \Delta\pi_{1_i})^{a_1} \, (\pi_{2_i}^{(o)} + \Delta\pi_{2_i})^{a_2} - \\ - (\pi_{1_i}^{(o)})^{a_1} \, (\pi_{2_i}^{(o)})^{a_2} \tag{3.26}$$

from which it follows that

$$d\phi_i = a_1 \frac{w_i^{(1)}}{\pi_{1_i}^{(o)} + \Delta\pi_{1_i}} d\Delta\pi_{1_i} + a_2 \frac{w_i^{(1)}}{\pi_{2_i}^{(o)} + \Delta\pi_{2_i}} d\Delta\pi_{2_i} \tag{3.27}$$

Write

$$\Delta\pi_{1_i} = \Sigma_j v_{1_j} \exp\left\{-\delta_1 c_{ij}\right\} \tag{3.28}$$

and

$$\Delta\pi_{2_i} = \Sigma_j v_{2_j} \exp\left\{-\delta_2 c_{ij}\right\} \tag{3.29}$$

in which $v_{1_j} = \Delta x_{1_j}$ and $v_{2_j} = \Delta x_{2_j}$

Then

$$d\phi_i = a_1 \frac{w_i^{(1)}}{\pi_{1i}^{(o)} + \Delta\pi_{1i}} \Sigma_j \exp\left\{-\delta_1 c_{ij}\right\} dv_{ij} + $$
$$+ a_2 \frac{w_i^{(1)}}{\pi_{2i}^{(o)} + \Delta\pi_{2i}} \Sigma_j \exp\left\{-\delta_2 c_{ij}\right\} dv_{2j} \qquad (3.30)$$

and

$$d\phi = \Sigma_i\, d\phi_i = a_1 \Sigma_i \frac{w_i^{(1)}}{\pi_{1i}^{(o)} + \Delta\pi_{1i}} \Sigma_j \exp\left\{-\delta_1 c_{ij}\right\} dv_{1j} + $$
$$+ a_2 \Sigma_i \frac{w_i^{(1)}}{\pi_{2i}^{(o)} + \Delta\pi_{2i}} \Sigma_j \exp\left\{-\delta_2 c_{ij}\right\} dv_{2j} \qquad (3.31)$$

An alternative approach would be to assume a 'government' welfare function (better to be called a government 'preference' function) being a function of the welfare levels of each individual region:

$$\phi = \phi(\phi_1, \ldots, \phi_n) \qquad (3.32)$$

so that

$$d\phi = \Sigma_i \phi_i' \, d\phi_i \qquad (3.33)$$

in which the weight ϕ_i' would be a declining function of the welfare level of the i-th region. [6] We will continue the argument with the simple assumption that $d\phi = \Sigma_i\, d\phi_i$, in which all weights are equal to one.

Now assume a budget restriction

$$\bar{y} = p_1 \Sigma_j v_{1j} + p_2 \Sigma_j v_{2j} \qquad (3.34)$$

in which $\bar{y}$ is the sum of money the government is willing and able to spend freely (i.e. after deduction of the running costs of the facilities already existing at the beginning of the year).

We now write

$$d\bar{y} = 0 = p_1 \Sigma_j dv_{1j} + p_2 \Sigma_j dv_{2j} \qquad (3.35)$$

and can derive the following first-order equilibrium conditions

$$\frac{\partial \phi}{\partial v_{1j}} = a_1 \Sigma_i \frac{w_i^{(1)} \exp\{-\delta_1 c_{ij}\}}{\pi_{1i}^{(o)} + \Delta\pi_{1i}} = \lambda p_1 \tag{3.36}$$

$$\frac{\partial \phi}{\partial v_{1k}} = a_1 \Sigma_i \frac{w_i^{(1)} \exp\{-\delta_1 c_{ik}\}}{\pi_{1i}^{(o)} + \Delta\pi_{1i}} = \lambda p_1 \tag{3.37}$$

$$\frac{\partial \phi}{\partial v_{2j}} = a_2 \Sigma_i \frac{w_i^{(1)} \exp\{-\delta_2 c_{ij}\}}{\pi_{2i}^{(o)} + \Delta\pi_{2i}} = \lambda p_2 \tag{3.38}$$

$$\frac{\partial \phi}{\partial v_{2k}} = a_2 \Sigma_i \frac{w_i^{(1)} \exp\{-\delta_2 c_{ik}\}}{\pi_{2i}^{(o)} + \Delta\pi_{2i}} = \lambda p_2 \tag{3.39}$$

The unknowns in these equations are the v_{1j}'s, the v_{2j}'s and λ, together 2n + 1 in number. The number of equations equals 2n plus the budget restriction, i.e. also 2n + 1. This means that the unknowns can in principle be determined as functions of the existing levels x_{1j} and x_{2j}, the distances c_{ij}, the two prices p_1 and p_2, and total government expenditures $\bar{y}$.

Although the equations look rather daunting, it may be remarked that the analysis results in a twofold optimisation procedure, viz., first, finding the optimum distribution of new facilities over the regions, and second finding the optimum expenditure on each facility. The first result may be written as

$$\frac{\partial \phi}{\partial v_{1j}} = \frac{\partial \phi}{\partial v_{1k}}, \text{ and alternatively } \frac{\partial \phi}{\partial v_{2j}} = \frac{\partial \phi}{\partial v_{2k}} \tag{3.40}$$

indicating that the marginal utility of the additional facility in j should equal the marginal utility of the additional facility in k.

The second may be written as:

$$\frac{\dfrac{\partial \phi}{\partial v_{1j}}}{\dfrac{\partial \phi}{\partial v_{2j}}} = \frac{p_1}{p_2}, \forall_j \tag{3.41}$$

a well-known formula, indicating that the ratio of the marginal utilities of facilities 1 and 2 should equal the ratio of their prices.

The solution of the system

The first-order conditions allow for the following solution procedure.
Write

$$z_{1i} = a_1 \frac{w_i}{\pi_{1i}} \tag{3.42}$$

$$z_{2i} = a_2 \frac{w_i}{\pi_{2i}} \tag{3.43}$$

Then equation (3.14) may be written as

$$\Sigma_i z_{1i} \exp\left\{-\delta_1 c_{ij}\right\} = \lambda p_1, \forall_j \tag{3.44}$$

or, in matrix notation

$$\mathbf{z}_1' \mathbf{D}_1 = \mathbf{i}' \lambda p_1 \tag{3.45}$$

in which $\mathbf{D}_1$ is the matrix of distance elements.
Similarly we write for the second facility

$$\mathbf{z}_2' \mathbf{D}_2 = \mathbf{i}' \lambda p_2 \tag{3.46}$$

From (3.23) and (3.24) we derive the values for $\mathbf{z}_1'$ and $\mathbf{z}_2'$

$$\mathbf{z}_1' = \mathbf{i}' \mathbf{D}_1^{-1} \lambda p_1 \tag{3.47}$$

and

$$\mathbf{z}_2' = \mathbf{i}' \mathbf{D}_2^{-1} \lambda p_2 \tag{3.48}$$

We proceed by rewriting – see also (3.25) – (3.42) and (3.43) as

$$\frac{w_i}{\pi_{1_i}} = \pi_{1_i}^{a_1 - 1} \pi_{2_i}^{a_2} = \frac{z_{1_i}}{a_1} \tag{3.49}$$

$$\frac{w_i}{\pi_{2_i}} = \pi_{1_i}^{a_1} \pi_{2_i}^{a_2 - 1} = \frac{z_{2_i}}{a_2} \tag{3.50}$$

or as

$$\ln \frac{z_{1_i}}{a_1} = (a_1 - 1) \ln \pi_{1_i} + a_2 \ln \pi_{2_i} \tag{3.51}$$

$$\ln \frac{z_{2_i}}{a_2} = a_1 \ln \pi_{1_i} + (a_2 - 1) \ln \pi_{2_i} \tag{3.52}$$

From these i sets of equations all π_{1_i} and π_{2_i} may be derived. Since the $\pi_{1_i}^{(o)}$ and $\pi_{2_i}^{(o)}$ are known, all $\Delta\pi_{1_i}$ and $\Delta\pi_{2_i}$ can be computed.

Since furthermore

$$v_1' D_1 = \Delta\pi'_1 \tag{3.53}$$

and

$$v_2' D_2 = \Delta\pi'_2 \tag{3.54}$$

we find

$$v_1' = \Delta\pi'_1 D_1^{-1} \tag{3.55}$$

and

$$v_2' = \Delta\pi'_2 D_2^{-1} \tag{3.56}$$

It appears that the values of v_1 and v_2 can be derived in a fairly straightforward way from the first-order equilibrium conditions. It should be noted, however, that the solution of the system has one important restriction viz., that all $v_i > 0$. This should, of course, be taken into account, by rewriting (3.36) through (3.39) as Kuhn-Tucker rather than as Lagrange conditions. [7]

Notes

[1] Co-authors of this chapter: L. van den Berg and C.H. Th. Vijverberg.
[2] The potentials are defined as weighted sums of x_j, regardless of the number of people using the facility. For comparison of potentials over regions we could bring in the number of users, but that aspect is left out for reasons of simplicity.
[3] As rankings are conserved under a monotone positive transformation the latter condition is too strong. However, as it is not relevant to the issue discussed in this section, no further attention will be paid to it.
[4] In fact, the λ's are marginal preference elasticities.
[5] In the foregoing sections no mention was made of the transport potential. The omission was deliberate: the influence of transport facilities on regional and urban welfare had been accounted for in each separate potential.
[6] Declining marginal preferences; once more, this assumption is not essential (see page 34).
[7] The equality-sign should be replaced by the inequality-sign $\leqslant$.

4. Potentials, distance and accessibility

Introduction

In the preceding chapter we already elaborated the concept of a potential. In this chapter we will go somewhat deeper into this subject and more particularly into the interrelations between potentials, accessibility and distance. Particularly the notion of a distance, which is an essential element in all concepts, deserves further attention. It is in spatial economics widely used but seldom with the precision that is necessary to separate the different sorts of distance we have in mind in a specific study. It is, however, obvious that distance in migration analysis is something quite different from distance in a study on commuting. In the former it is more of a proxy representing the 'strangeness' or psychological distance between the migrant's environment and the environment of other regions. In commuting, resulting from the spatial separation of living place and workplace, it is to be treated as generalised transportation costs with basic elements as money costs, time and comfort. In this chapter we will try to arrive at some definitions of 'distance' in different sorts of analysis. We will start with distance as generalised transportation costs and the relation of this concept to accessibility.

Distance and accessibility

A model can hardly be called a regional model if 'distance' in whatever form, does not play an essential role in it. The difference between general economics and spatial and regional economics is precisely that the distance factor is lacking in the former and plays an important role in the latter. It seems that the concept of a potential deals elegantly with this aspect in regional economics because it introduces in a systematic way the interrelation between regions.

Let us take a simple example to start with.

Write

$$r_{ij} = \frac{x_j \exp\{-ac_{ij}\}}{\Sigma_j x_j \exp\{-ac_{ij}\}} \quad r_i \qquad (4.1)$$

in which r_{ij} is the number of people seeking recreation in j and living in i. $r_i = \Sigma_j r_{ij}$ is the total number of people pursuing leisure activities in any j and living in i, c_{ij} represents the generalised costs of transportation from i to j as a weighted sum of money costs, time and efforts. x_j represents the attractiveness of j as a recreation area.

Assumption (4.1) implies that the total attraction exerted by a recreation area in j on inhabitants of i is determined by the attractiveness of the area itself (x_j) deflated by distance through the e-function. The fraction of the total number of people from i seeking recreation in j is then determined by the ratio of the attraction of j (corrected for distance) and the total attraction of all areas (including j).

We further assume

$$r_i = a_0 p_i (\pi_i)^{\beta_1} y_i^{\beta_2} \tag{4.2}$$

in which p_i = population in i

$\pi_i = \Sigma_j x_j \exp\{-ac_{ij}\}$

y_i = per capita income in i.

In this function π_i stands for the *recreation potential* of i. Obviously this potential, which (according to (4.2)) is one of the main explanatory factors of the total number of people at leisure, depends on the cost of transportation from i to all j-regions, as well as on all x_j.

The first conclusion we may draw, in so far as we are willing to accept the second hypothesis, is that an increase in all c_{ij}'s – which works out similarly as a corresponding increase in a (the 'resistance' against moving) – will decrease the potential and thus the total number of recreation-seeking people.

The second effect may be derived from the first hypothesis.

Write

$$r_{ik} = \frac{x_k \exp\{-ac_{ik}\}}{\Sigma_k x_k \exp\{-ac_{ik}\}} \tag{4.3}$$

then it follows from (4.1) and (4.3)

$$\frac{r_{ij}}{r_{ik}} = \frac{x_j}{x_k} \exp\{-a(c_{ij} - c_{ik})\} \tag{4.4}$$

This equation shows that a proportional increase in both c_{ij} and c_{ik}

increases the difference $c_{ij} - c_{ik}$ and consequently results in a relative decrease of r_{ij}. In other words, the increase in general transportation costs creates a preference for recreation areas closer by. This will result in a decrease in the average distance covered.

The final result is, then, that people will devote less time to leisure activities than before the rise in transportation costs, and cover an average distance shorter than before. Recreation activities will shrink in two ways, in the number of people and in average distance.

As we did in the preceding chapter, we will define the quality of life in a given region as a weighted product of all potentials that are important for the quality of life, such as recreation, housing environment, education, medical care, job opportunities, etc. All these elements can be expressed in potentials equally well as recreation. If educational possibilities are taken as an example, it seems justified to take into account not only the region in which people actually live, but also neighbouring regions, to a degree corresponding with their accessibility from their own region.

If we define quality of life in this way we must come to the conclusion that increased transportation costs influence the value of all potentials and consequently have on the one hand an adverse effect on the quality of life as a result of the decrease in the value of a number of potentials but on the other hand a favourable effect via the decrease in the volume of traffic. If less people move and move a smaller distance if they move, the result will be a decrease in the volume of traffic which, of course, should be judged favourably from a societal point of view.

Less traffic means fewer accidents, less air pollution and less space needed for infrastructure.

Lower potentials mean less recreation, less education, fewer job opportunities, lower environmental quality etc. Which of the two is more important is difficult to judge. The only thing we can say is that if the first factor is more important than the second, then the previous situation has been clearly sub-optimal, which means that taxation on transportation has been too low.

The reasoning concerning potentials that are essential for consumers can easily be extended to potentials important for producers. In order to keep the argument simple we will start with a sector that is completely demand-oriented, which means that its location is completely geared towards the location of effective demand.

We assume, similarly to what we did in the first part of this section,

$$d_i^p = \Sigma_j d_j \frac{\exp\{-\rho\, c_{ij}\}}{\phi_j} \qquad (4.5)$$

in which d^p_i = demand potential of region i

d_j = effective demand in j

ϕ_j = accessibility of region j (to be defined later).

(4.5) represents the weighted sum of demand in all regions j, deflated by communication costs. This concept will require further attention in a later section. ϕ_j represents the accessibility of region j from all regions. The easier j is accessible from all directions, the more difficult it will be for a producer in i to penetrate into this market. The value of ϕ_j can be found as follows.

$$\Sigma_i \, d^p_i = \Sigma_j d_j = d \qquad (4.6)$$

The sum of all potential demands has been defined here as the sum of all effective demands. This is, of course, not necessarily so. It is not difficult to show an example (e.g. relating to the labour market) where there is in some regions open demand (potential demand larger than effective demand) as well as open supply (potential supply larger than effective supply). In this example we will assume that the sum of all effective demands equals the sum of all potential demands so that there are no changes in stocks.

Consequently, we may write

$$d = \Sigma_i \Sigma_j \frac{d_j}{\phi_j} \exp\{-\rho c_{ij}\} = \Sigma_j \frac{d_j}{\phi_j} \Sigma_i \exp\{-\rho c_{ij}\} \qquad (4.7)$$

The obvious solution for ϕ_j is

$$\phi_j = \Sigma_i \exp\{-\rho c_{ij}\} \qquad (4.8)$$

The maximum value of ϕ_j is reached for all $c_{ij} = 0$. Then $\phi_j = n$ (number of regions). Its minimum value is reached for all $c_{ij} = \infty$. Then $\phi_j = 0$.

The *accessibility coefficient* a_j $(0 \leqslant a_j \leqslant 1)$ may now be written as

$$a_j = \frac{\phi_j}{n} = \frac{\Sigma_i \exp\{-\rho c_{ij}\}}{n} \qquad (4.9)$$

and (4.5) may be rewritten as

$$d^p_i = \Sigma_j d_j \frac{\exp\{-\rho c_{ij}\}}{\Sigma_i \exp\{-\rho c_{ij}\}} \qquad (4.10)$$

in which potential demand is defined as a realistic estimate of sales in all regions j of the industry located in i. Since ρ is sector specific, this equation obviously only holds for a well-defined sector.

Suppose now that c_{ij} increases. The first effect then is that the position of the i-producers on the j-market weakens.

We may easily derive that

$$\frac{\partial d_i^p}{\partial c_{ij}} = -\rho\, d_{ij}^p \left\{1 - \frac{d_{ij}^p}{d_j}\right\} \qquad (4.11)$$

in which d_{ij}^p represents the sales of i on the j-market. We thus find that apart from ρ the decrease of d_i^p is greater, the larger d_{ij}^p was and the better the relative position of others on the j-market was. In other words, the influence of an increase in generalised transportation costs is great if

a. the competitive position of i on the j-market was weak,

b. the sales in j were large in an absolute sense.

However, since we may assume that in this case $c_{ij} = c_{ji}$, there is a second influence of an increase in c_{ij}, viz. that the competitive position of j on the i-market weakens as the transportation costs c_{ji} will also increase.

We find

$$\frac{\partial d_i^p}{\partial c_{ji}} = \rho\, d_{ii}^p \cdot \frac{d_{ji}^p}{d_i} \qquad (4.12)$$

in other words, there is a positive effect on the position of the i-producers which is greater the larger the own sales in i were and the stronger the position of j in i was.

The total result equals the algebraical sum of the two effects

$$\frac{\partial d_i^p}{\partial c_{ij}} + \frac{\partial d_i^p}{\partial c_{ji}} = \frac{-\rho\, d_j \exp\{-\rho\, c_{ij}\}}{\Sigma_i} \cdot \frac{\Sigma_i \exp\{-\rho\, c_{ij}\} - \exp\{-\rho\, c_{ij}\}}{\Sigma_i} +$$

$$+ \frac{\rho\, d_i \exp\{-\rho\, c_{ij}\}}{\Sigma_j} \cdot \frac{1}{\Sigma_j} = -\rho \exp\{-\rho\, c_{ij}\} \left\{ \frac{d_j}{\Sigma_i} \cdot \frac{\Sigma_i \exp\{-\rho\, c_{ij}\} - \exp\{-\rho\, c_{ij}\}}{\Sigma_i} + \right.$$

$$\left. + \frac{d_i}{\Sigma_j} \cdot \frac{1}{\Sigma_j} \right\} \qquad (4.13)$$

where

Σ_i stands for $\Sigma_i \exp\left\{-\rho\, c_{ij}\right\}$

and

Σ_j stands for $\Sigma_j \exp\left\{-\rho\, c_{ij}\right\}$

This expression is negative if

$$\frac{d_j}{d_i}\left(\frac{\Sigma_j}{\Sigma_i}\right)^2 > \frac{1}{\Sigma_i - \exp\left\{-\rho\, c_{ij}\right\}} \qquad (4.14)$$

which is relatively small compared to one.

Since we may write

$$\frac{d_j}{d_i}\left(\frac{\Sigma_j}{\Sigma_i}\right)^2 = \frac{d_j}{d_i}\left(\frac{a_j}{a_i}\right)^2 \qquad (4.15)$$

the results imply that the effect will be negative even if the j-market is relatively small compared to the i-market and relatively inaccessible compared to the i-market.

The main conclusion that can be drawn from (4.13) is that smaller (small d_i) markets and medium-sized will lose and very large markets will gain when the level of transportation costs increases, in other words, regional policy directed, as it generally is, towards the development of smaller markets will become more difficult. With the rise in transportation costs the sizes of the relevant regions will decrease and the regional economies will move to a certain extent in the direction of more autarky with less interregional trade and a stronger concentration on their own market.

Evidently, a similar reasoning can be set up for the inputs needed for production. As these can also be expressed as potentials, each region will suffer from a general increase in transportation costs lowering the accessibility of inputs. However, here too, larger, diversified markets will suffer considerably less than smaller markets with a less diversified structure and higher dependence on importation of inputs needed for production.

Both trends show a heavier dependence of local production on local markets and inputs locally available as a result of a general decrease in the values of potentials. It appears that the effects on production households are similar to those on consumer households. The same holds for the labour market. Not only will rising transportation costs decrease the job opportunities for workers, they will just as well influence negatively the availability of workers of the required skill-levels for production households. All potentials will shrink, with both favourable and unfavourable effects. Whether the balance will ultimately be positive or negative will be considered in a later section.

Fuel prices and communication costs

As already indicated, transportation costs for households should be generally seen as so-called generalised transportation costs. They represent the weighted sum of money costs, time and effort, and the risks of bridging distance.

In order to simplify the argument, we will assume that generalised transportation costs consist of three elements, viz., money costs of transportation, which are assumed to be proportional to distance, time costs, and risk factors, both also proportional to distance.

We then find

$$c_{ij} = (m + r + \frac{\lambda}{v} y)\, d_{ij} \qquad (4.16)$$

in which m = money costs per unit of distance
r = risk per unit of distance
v = average speed
λy = value of the time unit.

Now an essential part of m is fuel or energy costs. So, we may write

$$\frac{e_{ij}}{c_{ij}} = \frac{e}{m + r + \frac{\lambda}{v} y} \qquad (4.17)$$

in which e_{ij} = total energy costs of the journey in j
e = energy costs per unit of distance.

In the foregoing we have considered the consequences of a rise in transportation costs. Here it becomes clear that a rise in the energy share of transportation costs may be the result of

a. a rise in fuel prices (e)
b. a decrease in risk (r)
c. an increase in speed (v)
d. a decrease in income (y)

We found earlier that a rise in fuel costs will lead to a decrease in traffic volume owing to the resulting increased costs of bridging distances. This implies that to the extent that this is true, all above mentioned effects will occur simultaneously, since lower traffic volumes will lead to smaller risks and increases in speed, while a lower level of income results from the exogenously determined heavy increase in energy prices.

It appears that rising fuel prices will result in higher transportation costs together with a heavy increase of the relative share of energy prices in these higher costs.

The situation for production households is somewhat but not much different. The costs of communication are here in many instances more important than actual transportation costs. In fact there are three elements that should be taken into account in the demand- and supply-potentials, viz., (a) transportation costs of goods, (b) costs of face-to-face contacts, (c) costs of telecommunication.

More than consumption households, production households will be able to substitute face-to-face contacts by telecommunication, although it cannot be denied that the quality of this kind of communication is usually inferior to that of face-to-face contacts. The impact of the rise in energy prices will be highest in sectors in which transportation of goods is relatively important. This will in general be basic industries. Telecommunication is supposedly most important in the business-service sector, which will be affected least, at least directly, by the increase in energy prices.

The foregoing arguments may show that the rise in fuel prices may have considerable influence through the effect they have on transportation costs. Of course, this will not be the only effect. Energy raw materials are also used in other activities than transportation, notably for equipment units and as raw materials in chemical production. This means that the price rise will spread over all sectors via the accumulation of energy-price effects as well as via the accumulation of transportation-price effects.

This development will work out differently for different sectors of the economy and will consequently have different effects upon different regions.

It is, however, the object of this section to concentrate on the general effects in a spatial context, rather than on the specific influences in specific regions, although the latter no doubt deserve extensive attention. With respect to the specific spatial effects we will now try to draw some general conclusions from the foregoing argument.

Some general conclusions

Schematically, the argument developed so far with regard to consumers' households can be presented as follows.

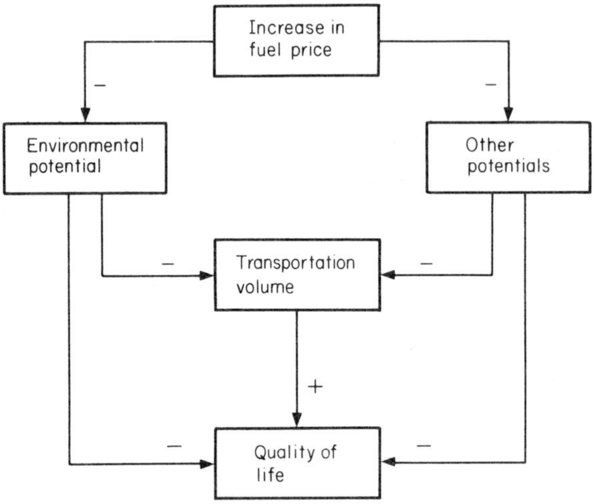

The increase in fuel prices will lead, through the effects shown before, to a general decrease in the value of potentials. Although these potentials could be taken together, it seems worthwhile to mention the environmental potential separately to indicate that the effects of the fuel price on this potential is twofold.

The decrease in potentials will eventually lead to a decrease in the volume of transportation. The decrease of the potentials itself can be considered as a negative contribution to the quality of life, the decrease in the volume of transportation, however, as a positive contribution.

For production households the situation is as follows.

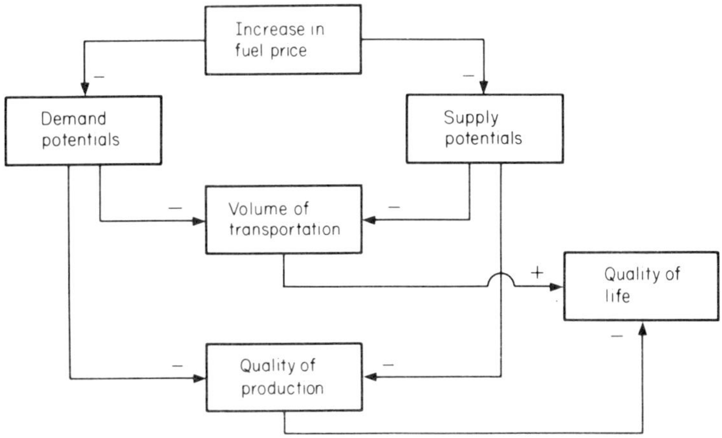

The increase in the energy prices causes a decrease in demand potentials as well as in supply potentials as a result of the decreased accessibility of markets for outputs as well as for inputs. Both decreases lead eventually to a decrease in the volume of transportation which, as it did in the former scheme, favourably influences the quality of life.

The quality of production in general, however, will decrease. It will suffer from the diminished accessibility of both markets and inputs as well as from the more limited availability of qualified workers as a result of the decreased commuting distances implying smaller areas from which workers can be attracted. This influences negatively the quality of life, since the quality of goods and services produced is an element of the quality of life.

Taking both groups of effects together we find

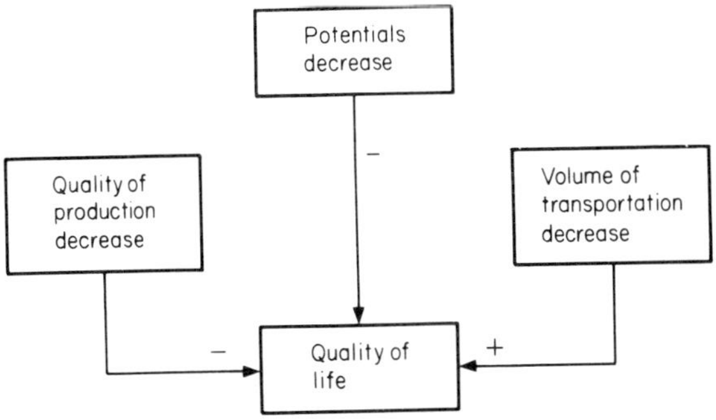

This diagram shows that, in fact, the only advantage in spatial development is the decrease in the volume of traffic. All other effects are negative.

As far as we may assume that the fuel prices in the new situation represent better than before the real scarcity of energy raw materials, we should of course accept the new situation as being better than the original one. Many of us will certainly also welcome the decrease in the volume of traffic as a contribution to the quality of life, even taking into consideration the sacrifices to be made for it in other fields.

We should not forget, however, that so far we have only looked at the changes in the values of the potentials, and have not yet considered the impact of these changes on the location of households of both consumers and producers, which will be done in a later chapter. Under the process of suburbanisation workers might have chosen residences rather far from their working place since transportation costs being relatively unimportant in their considerations about their residence. If these costs increase considerably, their opinion might change and they might be inclined to locate closer to their work or to look for another job closer by. Firms might find themselves confronted with heavily increasing expenses for the transportation of their inputs and their final products and might consequently consider a change in location, particularly if they are located in relatively isolated areas, where in most cases they are already leading a marginal existence anyway.

Social distance

In the theory of social communication it is assumed that the intensity of communication between different social and/or economic groups is determined *inter alia* by the social distance between these groups. Social distance is supposed to indicate a degree of separation, which acts as a barrier, that is experienced by the members of the groups.

However, in practical research concerning communication flows (e.g. traffic, migration, telephone calls) one often observes a transition from social distances between *groups* to psychological distances between *regions* these groups live in. In this chapter a similar procedure is adopted by the authors.

In empirical applications social distance is often replaced by physical distance because of the fact that until now one did not succeed in quantifying this not very well defined concept. Certainly, physical distance is a very important element in social distance, but social distance should not just be treated as a function of distance only.

The social distance between two regions is in this view not only dependent upon the distance between those regions, but also on the composition of both regions as far as social and/or economic groups are concerned.

In this chapter an effort is made to give an operational definition of social distance.

Communication intensities as a measure for social distance

Suppose the following deterministic expression

$$v_{ij}^{kl} = a^{kl} s_i^k s_j^l \exp\left\{-\beta d_{ij}\right\} \quad \begin{matrix} k, l = 1, \ldots, s; \\ i, j = 1, \ldots, r \end{matrix} \tag{4.18}$$

in which v_{ij}^{kl} is the size of communication per unit of time between group k in i and group l in j, in which the initiative is taken by members of group k in i.

a^{kl} is the intensity of communication between one member of group k and one member of group l if the initiative of the communication is taken by the first and if the distance between the two members were zero.

s_i^k is the size of group k in i

s_j^l is the size of group l in j

d_{ij} is the distance between region i and region j.

Now suppose that the maximum value that a^{kl} can assume is a^*. Then (4.18) may be written as

$$v_{ij}^{kl} = a^* s_i^k s_j^l \gamma^{kl} \exp\left\{-\beta d_{ij}\right\} \tag{4.19}$$

in which

$$\gamma^{kl} = \frac{a^{kl}}{a^*} \tag{4.20}$$

Now write

$$\begin{aligned} v_{ij}^{kl} &= a^* s_i^k s_j^l \exp\left\{-\beta d_{ij} + \ln \gamma^{kl}\right\} \\ &= a^* s_i^k s_j^l \exp\left\{-\beta \left(d_{ij} - \frac{1}{\beta} \ln \gamma^{kl}\right)\right\} \end{aligned} \tag{4.21}$$

Define Ψ_{ij}^{kl} as

$$\Psi_{ij}^{kl} = d_{ij} - \frac{1}{\beta} \ln \gamma^{kl} \qquad (4.22)$$

then (4.21) may be written as

$$v_{ij}^{kl} = a^* \, s_i^k \, s_j^l \, \exp\left\{-\beta \, \Psi_{ij}^{kl}\right\} \qquad (4.23)$$

Ψ_{ij}^{kl} will be called the social distance between group k in i and group l in j. If the distance $d_{ij} = 0$ this social distance is equal to $-\frac{1}{\beta} \ln \gamma^{kl}$ which is a non-negative number since $0 < \gamma^{kl} < 1$. The dimension of this expression is 'distance' since the dimension of $\frac{1}{\beta}$ is 'distance' and $\ln \gamma^{kl}$ is dimensionless.

v_{ij} is given by

$$v_{ij} = \Sigma_k \, \Sigma_l \, v_{ij}^{kl} \qquad (4.24)$$

Substitution of (4.18) into (4.24) results in

$$v_{ij} = \Sigma_k \, \Sigma_l \, a^{kl} \, s_i^k \, s_j^l \, \exp\left\{-\beta d_{ij}\right\} \qquad (4.25)$$

From (4.25) we find

$$v_{ij} \exp\left\{\beta d_{ij}\right\} = \Sigma_k \, \Sigma_l \, a^{kl} \, s_i^k \, s_j^l \qquad (4.26)$$

Now write **V** for the r x r matrix with typical element $v_{ij} \exp\left\{\beta d_{ij}\right\}$
S for the r x s matrix with typical element s_i^k
A for the s x s matrix with typical element a^{kl}
Γ for the s x s matrix with typical element γ^{kl},

then

$$\mathbf{V} = \mathbf{S\,A\,S'} \qquad (4.27)$$

or

$$\mathbf{V} = a^* \, \mathbf{S\,\Gamma\,S'} \qquad (4.28)$$

If s_i represents the total size of the population of region i and f_i^k is defined as

$$f_i^k = \frac{s_i^k}{s_i} \qquad (4.29)$$

then (4.26) may be written – see also (4.20) – as

$$v_{ij} \exp\{\beta d_{ij}\} = a^* \Sigma_k \Sigma_l s_i f_i^k \gamma^{kl} f_j^l s_j$$

$$= a^* s_i s_j \Sigma_k \Sigma_l f_i^k \gamma^{kl} f_j^l \quad (4.30)$$

If $\hat{s}$ represents the diagonal matrix of the total sizes of the population of all r regions and F the r x s matrix of the fractions of all groups in all regions, then (4.28) can be written as follows

$$V = a^* \hat{s} F \Gamma F' \hat{s} \quad (4.31)$$

Now define

$$F \Gamma F' = \bar{\Gamma} \quad (4.32)$$

The typical element $\bar{\gamma}_{ij}$ of $\bar{\Gamma}$ is the average of all γ^{kl} weighted with the socio-economic structures in both i and j; see (4.30).

Combination of (4.32) and (4.31) gives

$$V = a^* \hat{s} \bar{\Gamma} \hat{s} \quad (4.33)$$

The typical element of (4.33) equals

$$v_{ij} \exp\{\beta d_{ij}\} = a^* s_i s_j \bar{\gamma}_{ij} \quad (4.34)$$

Similar to (4.18) up to (4.22) we may rewrite (4.34) as follows

$$v_{ij} = a^* s_i s_j \bar{\gamma}_{ij} \exp\{-\beta d_{ij}\}$$

$$= a^* s_i s_j \exp\{-\beta d_{ij} + \ln \bar{\gamma}_{ij}\}$$

$$= a^* s_i s_j \exp\{-\beta (d_{ij} - \frac{1}{\beta} \ln \bar{\gamma}_{ij})\} \quad (4.35)$$

Define

$$\Psi_{ij} = d_{ij} - \frac{1}{\beta} \ln \bar{\gamma}_{ij} \quad (4.36)$$

then

$$v_{ij} = a^* s_i s_j \exp\{-\beta \Psi_{ij}\} \quad (4.37)$$

Ψ_{ij} represents the social distance between region i and region j. It should be remarked that $\Psi_{ij} \neq \Psi_{ji}$.

Substitution of (4.23) into (4.24) results in

$$v_{ij} = a^* \Sigma_k \Sigma_l \; s_i^k s_j^l \exp\left\{-\beta \Psi_{ij}^{kl}\right\} \tag{4.38}$$

Using (4.29) we find

$$v_{ij} = a^* s_i s_j \Sigma_k \Sigma_l \; f_i^k f_j^l \exp\left\{-\beta \Psi_{ij}^{kl}\right\} \tag{4.39}$$

Combination of (4.37) and (4.39) gives

$$\exp\left\{-\beta \Psi_{ij}\right\} = \Sigma_k \Sigma_l \; f_i^k f_j^l \exp\left\{-\beta \Psi_{ij}^{kl}\right\} \tag{4.40}$$

and also

$$\exp\left\{-\beta \Psi_{ij}\right\} = \exp\left\{-\beta\right\} \Sigma_k \Sigma_l \; f_i^k f_j^l \exp\left\{-\beta (\Psi_{ij}^{kl} - 1)\right\}$$

$$= \exp\left\{-\beta \left\{1 - \frac{1}{\beta} \ln \Sigma_k \Sigma_l \; f_i^k f_j^l \exp\left\{-\beta(\Psi_{ij}^{kl} - 1)\right\}\right\}\right\} \tag{4.41}$$

From (4.41) we find

$$\Psi_{ij} = 1 - \frac{1}{\beta} \ln \Sigma_k \Sigma_l \; f_i^k f_j^l \exp\left\{-\beta (\Psi_{ij}^{kl} - 1)\right\}$$

$$= -\frac{1}{\beta} \ln \Sigma_k \Sigma_l \; f_i^k f_j^l \exp\left\{-\beta \Psi_{ij}^{kl}\right\} \tag{4.42}$$

Expression (4.42) shows the relation between the social distance of region i and j and the social distances between all social or economic groups in these regions.

An alternative approach; regional communication intensities

An alternative approach is found if it is assumed that

$$a^{kl} = a^k a^l \tag{4.43}$$

This assumption means that the intensity of communication of a member of group k is independent of the group to which the other communicator belongs; in other words the intensity of communication is independent of its social direction. Per unit of time a^k messages are sent from one member of group k to one member of group l, each message resulting in a^l response-messages from the member of group 1.

(4.43) means that we may write

$$A = aa' \qquad (4.44)$$

in which $a' = \{a^1, \ldots, a^s\}$, the row vector of communication intensities per group.

From (4.29) we have

$$S = \hat{s}\, F \qquad (4.45)$$

Now we define the column vector

$$\gamma = \frac{1}{a^*}\, a \qquad (4.46)$$

where a^* is the maximum value of a^k and where $r' = \{\gamma^1, \ldots, \gamma^s\}$.

Then (4.44) may be written as

$$A = a^{*2}\, \gamma\gamma' \qquad (4.47)$$

Substitution of (4.45) and (4.47) in (4.27) results in

$$V = a^{*2}\, \hat{s}\, F\, \gamma\gamma'\, F'\, \hat{s} \qquad (4.48)$$

Now write

$$\gamma^* = F\, \gamma \qquad (4.49)$$

in which γ^* represents the vector of communication coefficients per region and where $\gamma^* = \{\gamma^*, \ldots, r^*\}$

Expression (4.48) now becomes

$$V = a^{*2}\, \hat{s}\, \gamma^*\, \gamma^{*\prime}\, \hat{s} \qquad (4.50)$$

The typical element of (4.50) is

$$v_{ij} \exp\{\beta d_{ij}\} = a^{*2}\, s_i\, s_j\, r_i^*\, r_j^* \qquad (4.51)$$

From (4.51) we have

$$v_{ij} = a^{*2}\, s_i\, s_j \exp\{-\beta d_{ij} + \ln r_i^* + \ln r_j^*\}$$

$$= a^{*2}\, s_i\, s_j \exp\{-\beta\,(d_{ij} - \frac{1}{\beta} \ln r_i^* - \frac{1}{\beta} \ln r_j^*)\} \qquad (4.52)$$

The social distance between region i and j is defined in the alternative approach as

$$\Psi_{ij}^* = d_{ij} - \frac{1}{\beta} \ln r_i^* - \frac{1}{\beta} \ln r_j^* \qquad (4.53)$$

Obviously in this case $\Psi^*_{ij} = \Psi^*_{ji}$.

Finally, combining (4.52) and (4.53)

$$v_{ij} = a^{*2} s_i s_j \exp\left\{-\beta\Psi^*_{ij}\right\} \tag{4.54}$$

Similar to (4.22) and (4.23) we may write

$$\Psi^{*kl}_{ij} = d_{ij} - \frac{1}{\beta} \ln \gamma^k - \frac{1}{\beta} \ln \gamma^1 \tag{4.55}$$

and

$$v^{kl}_{ij} = a^{*2} s^k_i s^l_j \exp\left\{-\beta \Psi^{*kl}_{ij}\right\} \tag{4.56}$$

Substitution of (4.56) and (4.24) gives

$$v_{ij} = a^{*2} \Sigma_k \Sigma_l s^k_i s^l_j \exp\left\{-\beta \Psi^{*kl}_{ij}\right\} \tag{4.57}$$

From (4.54), (4.57) and (4.29) we conclude

$$\exp\left\{-\beta \Psi^*_{ij}\right\} = \Sigma_k \Sigma_l f^k_i f^l_j \exp\left\{-\beta \Psi^{*kl}_{ij}\right\} \tag{4.58}$$

From (4.58) one derives – see also (4.41) and (4.42) –

$$\Psi^*_{ij} = -\frac{1}{\beta} \ln \Sigma_k \Sigma_l f^k_i f^l_j \exp\left\{-\beta\Psi^{*kl}_{ij}\right\} \tag{4.59}$$

Power function or e-function

Suppose

$$v^{kl}_{ij} = a^{kl} s^k_i s^l_j d^{-\beta}_{ij} \qquad \begin{array}{l} k, l = 1, \ldots, s; \\ i, j = 1, \ldots, r \end{array} \tag{4.60}$$

It may be remarked that if the influence of distance is represented by a power function, social distance between two groups becomes zero if physical distance is zero. This may be shown as follows.

Using (4.20) we have

$$v_{ij}^{kl} = \alpha^* s_i^k s_j^l \gamma^{kl} d_{ij}^{-\beta}$$
$$= \alpha^* s_i^k s_j^l \left[d_{ij} (\gamma^{kl})^{-1/\beta}\right]^{-\beta}$$
$$= \alpha^* s_i^k s_j^l \Psi_{ij}^{-\beta} \qquad (4.61)$$

in which

$$\Psi_{ij} = d_{ij} (\gamma^{kl})^{-1/\beta} \qquad (4.62)$$

This expression implies that there exists no social distance between different social groups if the physical distance between the members of those groups is zero. Since this result is unacceptable, the hypothesis that the distance influence may be represented by a power function, is rejected.

More general definitions

The concept of a distance

If we combine the examples that have been presented in the foregoing sections we may arrive at a general definition of distance and resistance it exerts on spatial phenomena. Obviously the elements of distance that play a role are

1. Money costs needed to pay for the bridging of the physical distance. Physical distance is here defined as the straight line distance between origin and destination so that necessary deviations are incorporated in the price per distance unit.
2. Time that has to be sacrificed for bridging the distance.
3. Social distance which particularly plays a role in communication but also might be of importance in migration (different languages, different religions, etc.).
4. The 'strangeness' of the region of destination as function of physical distance. It should be remarked here that physical distance might also be large in case a region is very much isolated because of topological reasons.
5. The risk of accidents involved in travelling.

We now may write for the probability that a certain interaction will take place

$$p_{ij}^{kl} = \exp\left\{-\beta_1 t d_{ij} - \beta_2 \frac{d_{ij}}{V} - \delta^{kl} - \beta_3 d_{ij} - \beta_4 \rho d_{ij}\right\} \tag{4.63}$$

in which t = transportation costs per unit of distance

v = average speed

δ^{kl} = social distance

d_{ij} = physical distance

ρ = risk in bridging the distance between i and j per unit of distance

Obviously the importance of time is related to the marginal utility of income. This point will be discussed in a later section.

(4.63) is a general expression that holds also for communication although the risk factor there may be put at zero. Since communication takes to a large extent place on an individual basis, δ^{kl} usually will play a considerable role. However, also in migration it may happen that δ^{kl} is important, particularly if the social structure of the region of origin is very different from that in the region of destination.

In commuting the first two factors are of considerable importance but also might the strangeness of the region of destination (if the residential location is given and a job is selected) play a role in the selection of the job. Since commuting takes place over relatively small distances, this factor will, however, not be very important.

In the use of amenities (e.g. shopping, recreation) both money and time costs play a role. Social distance may be neglected. The strangeness of the region of destination usually has a temporary influence. It demonstrates itself in shopping when people have migrated from one region to another and maintain for some time the habit of shopping in their former living place since this place is better known to them. After a period of a year, however, they have become accustomed to the new circumstances and will behave accordingly. The risk for accidents plays an important role for young children on their way to the kindergarten or primary school. The risk may be diminished by providing schoolbuses.

In this way it is possible to express some a priori expectations about the importance of the different factors for different activities. We summarised some of them in the following table.

Activity	Importance of resistance factors in spatial activities				
	Money	Time	Social distance	Strangeness (physical distance)	Risk
Commuting	*	*	—	—	—
Communication	*	*	*	*	—
Use of amenities	—	*	—	—	(*)
Migration	—	—	(*)	*	—

The concepts of accessibility and potentials

Although the concept of accessibility was already treated in chapter 4, section 'Distance and accessibility', p.48, it will be useful to treat the subject in a somewhat more general way. Remembering equation (4.9) we start with the most simple definition of accessibility

$$a_i = \frac{1}{n} \Sigma_j \exp \left\{-\alpha d_{ji}\right\} \tag{4.64}$$

Here accessibility is defined as the unweighted sum of all resistance factors over the regions under consideration.

The value of (4.64) lies between zero and unity and depends on the 'distance' matrix as well as the value of α. The first dependence is obvious, the smaller the distances, the more accessible a region becomes, whatever the influence of α. α itself poses somewhat more problems. Let us consider the elements represented by α as they can be derived from (4.63).

We then find

$$\alpha = \beta_1 t + \frac{\beta_2}{v} + \beta_3 + \beta_4 \rho \tag{4.65}$$

Consisting of the elements money costs, time, 'strangeness' and risk per unit of physical distance as well as with the weights attached to these elements β_1 through β_4. Now, given the explaining variables (t, $\frac{1}{v}$ and ρ) it is precisely the coefficients β_1 through β_4 that determine the value of α. Thus we may apply the information given in the table above to see what values may be expected for α in different activities. The extremely important conclusion then is that α is activity-specific, in other words that for some activities α might be low, which is often the case if the frequency with which the activity takes place is low (holiday, migration) while for other activities it might be very high (shopping or daily goods, commuting) or average (weekend recreation).

The foregoing implies that, *if a is activity-specific, accessibility is also activity-specific.* A region might be highly accessible for one activity and inaccessible for another.

Let us give a simple example. We take 2 values for a, $a = 0.05$ and $a = 0.5$ respectively. The distances from 4 regions to the fifth one are given as 5, 10, 20 and 25. It then may be calculated that the a for the fifth region (if the distance to their own region is put at zero) is 0.72 in case $a = 0.05$ but only 0.28 in case $a = 0.5$.

So far we have considered the accessibility of a given region from all regions. We also could study the accessibility of all regions from one region, which is the reverse case. We may define this as

$$b_i = \frac{1}{n} \Sigma_j \exp \left\{-a d_{ij}\right\} \tag{4.66}$$

which is identical to a_i if $d_{ij} = d_{ji}$.

Written in matrix notation we obtain as definitions

$$\mathbf{a} = \frac{1}{n} \mathbf{A}'\mathbf{i} \tag{4.67}$$

and

$$\mathbf{b} = \frac{1}{n} \mathbf{A}\mathbf{i} \tag{4.68}$$

in which the typical element of the square matrix $\mathbf{A}$ is $\exp \left\{-a d_{ij}\right\}$. Both expressions are identical for $\mathbf{A} = \mathbf{A}'$.

It will be evident that in many cases $\mathbf{A} \neq \mathbf{A}'$. In commuting, for instance, city centres may be highly inaccessible from other parts of the city during the rush hour while during the same rush hour (which is basically a one direction phenomenon) other parts of the city are easily accessible from the centre.

Our first conclusion thus is that it is very important to distinguish between these two sorts of accessibilities. We will therefore call a (the accessibility of a given region from all regions) the *inward accessibility* and b (the accessibility of all regions from a given region) the *outward accessibility.* The first concept is of prior importance to those who are dependent on inflows of people (industrial firms demanding labour, shops demanding shoppers, recreation areas demanding recreants, etc.) while the outward accessibility is the important thing for people living in a given region and wanting to move regularly to other regions for the same purposes (work, recreation, shopping, etc.). The more all regions are accessible from their own region, the better off they are.

It is here the point to introduce a new factor. If we assume, like we did in the preceding paragraph, the inhabitant in a given region is interested in the easiness with which he is able to visit other places in his own as well as other regions, he is usually interested for a specific reason. Let us consider shopping. If considered in this context the potential shopper is not interested in the accessibility of another region from the place where he lives if the other region does not offer him facilities in the field of shopping. He will be the more interested in the other regions the better the shopping facilities there are. The same holds for all other activities. A region might be interesting for him if there are many job opportunities or good schools or good recreation facilities. He will be more interested in the accessibility of a region where those facilities are readily available than in a region where they are absent or only available in marginal quantities. This implies that we will have to introduce a system of region-specific weights in both definitions of accessibility.

We now rewrite (4.66) as

$$b_i = \frac{\Sigma_j x_j \exp\left\{-ad_{ij}\right\}}{\Sigma x_j} = \Sigma_j \frac{x_j}{n\bar{x}} \exp\left\{-ad_{ij}\right\}$$

$$= \Sigma_j \phi_j \exp\left\{-ad_{ij}\right\} \qquad (4.69)$$

and similarly

$$a_i = \frac{\Sigma_j z_j \exp\left\{-ad_{ji}\right\}}{\Sigma z_j} = \Sigma_j \frac{z_j}{n\bar{z}} \exp\left\{-ad_{ji}\right\}$$

$$= \Sigma_j \psi_j \exp\left\{-ad_{ji}\right\} \qquad (4.70)$$

or, in matrix notation

$$\mathbf{a} = \mathbf{A}' \hat{\Psi} \mathrm{i} \qquad (4.71)$$

and

$$\mathbf{b} = \mathbf{A} \hat{\Phi} \mathrm{i} \qquad (4.72)$$

It is obvious that x_j and z_j are different things. The shopper is interested in the accessibility of shops (x_j) while the shopkeepers are interested in the accessibility of shoppers (z_j). We thus not only may find a difference in the value of **a** and **b** due to the fact that $A \neq A'$ but also because $\hat{\Psi} \neq \hat{\Phi}$.

In proceeding in this direction we come very close to the concept of potentials. In fact, we may rewrite (4.69) as

$$b_i = \frac{1}{n\bar{x}} \pi_i^x \quad (4.73)$$

and

$$a_i = \frac{1}{n\bar{z}} \pi_i^z \quad (4.74)$$

(4.71) and (4.72) then become

$$a = \frac{1}{n\bar{x}} A' \hat{X} i = \frac{1}{n\bar{x}} \pi^x \quad (4.75)$$

$$b = \frac{1}{n\bar{z}} A \hat{Z} i = \frac{1}{n\bar{z}} \pi^z \quad (4.76)$$

which means that the accessibility indices are proportional with their corresponding potentials. This demonstrates the very close connections between the two concepts.

It appears that the concept of inward accessibility is closely related with producers' potentials (π^x) while the concept of outward accessibility relates directly to consumers' potentials (π^z). More general terms for consumers' and producers' potentials might be buyers' potentials and sellers' potentials.

Mobility

It seems useful to dedicate a few lines to the concept of mobility which is implicitly used in the foregoing analysis. Let us take $\frac{1}{\alpha}$ (see (4.65)) as a measure of mobility (μ).

We then find

$$\alpha = \frac{1}{\mu} = \beta_1 t + \frac{\beta_2}{v} + \beta_3 + \beta_4 \rho \quad (4.77)$$

Mobility is then defined as a coefficient that depends upon the same variables that determine accessibility and the potentials. This seems quite logical. They are decisive for the size of the resistance exerted per unit distance of the four factors represented in (4.77). It even seems logical to speak of specific types of mobility, each type related to one of the elements money, time, 'strangeness' and risk. For each spatial activity the β's, as we have seen, may have different values, resulting in different values for the mobility coefficient.

Defined in this way mobility is a function of both the coefficients β as well as of the values of the resistance factors themselves. Which of course can only be defined in three of the four cases.

An alternative approach could have been to distinguish four mobilities defined in the following manner:

$$\frac{1}{\mu} = \beta_1$$

$$\frac{1}{\mu} = \beta_2$$

$$\frac{1}{\mu} = \beta_3$$

$$\frac{1}{\mu} = \beta_4 \tag{4.78}$$

μ_1 is then the mobility with respect to the money expenses per unit distance of the trip and μ_2 the mobility with respect to the time needed per unit distance. μ_3 is the mobility with respect to the strangeness of the region of destination and μ_4 that with respect to risk per unit distance.

Although it is largely a matter of personal taste, it seems that the second approach is to be preferred since it separates the influence of the variables from those of the coefficients and mobilities are consequently defined more precisely.

It may, very generally, be said that one might expect that β_1 is negatively related to income and β_2 is positively related to income. The value of β_3 will diminish with increased educational level as well as with the quality of the available information and experience while β_4 will be high for both very young and very old people.

References

van den Berg, L., Klaassen, L.H. and Vijverberg, C.H.Th., *Evaluation of Governmental Welfare Policy by means of a Social Welfare Function,* FEER 1975/7, Netherlands Economic Institute, Rotterdam, 1975.

de V. Graaff, J.,*Theoretical Welfare Economics*, Cambridge University Press, 1967.

ter Heide, H., *Binnenlandse migratie in Nederland*, 's-Gravenhage, 1965.

Hordijk, L., Mastenbroek, P. and Paelinck, J.H.P., *Estimation of relative preference elasticities, A simulation approach*, FEER 1974/9, Netherlands Economic Institute, Rotterdam, 1974.

Isard, W., *Methods of Regional Analysis*, MIT Press, New York, 1960.

Klaassen, L.H. and Verster, A.C.P., *SPAMO I – Een ruimtelijk model*, Nederlands Economisch Instituut, Rotterdam, 1973/74.

Nijkamp, P., Theorie en toepassing van graviteits- en entropiemodellen in ruimtelijke systemen, *Research Memorandum no.11*, Vrije Universiteit, Amsterdam, 1974.

Nijkamp, P., Operational Determination of Collective Preference Parameters: A Survey of revealed preference methods in collective decision-making, *Research Memorandum no.17*, Free University, Amsterdam, 1975.

Nijkamp, P., Determination of implicit social preference functions, *Report 7010*, Econometric Institute, Netherlands School of Economics, Rotterdam, 1970.

Paelinck, J.H.P., Modèles de Politique Economique Multirégionale Basés sur l'Analyse d'Attraction, *L'Actualité Economique*, octobre-décembre, 1973.

Schouten, C.W., Binnenlandse migratie en regionale politiek, *Economisch Statistische Berichten*, 22-1-1975, pp 76-81.(a)

Schouten, C.W., De regionale doorlichting van Nederland; achtergronden en uitkomsten, *Economisch Statistische Berichten*, 15-1-1975, pp 52-6. (b)

Wilson, A.G., A family of spatial interaction models, and associated developments, *Environment and Planning*, vol.3, (1971), pp 1-32.

PART II

THEORETICAL SPATIAL MODELS

5. Residential location and social infrastructure models (SPAMOS)

Introduction: the general structure of the models

The distribution function

Residential location is, apart from the characteristics of the population (age groups, professional status, income, etc.) and the job opportunities (to which we will come later) determined by the quality and quantity of amenities in the region, housing quality and environmental qualities. In this chapter we will dedicate ourselves to the amenities and present first a general structure for the models to be used in shopping, education, recreation, etc.

We write

$$n_{ij} = \frac{x_j e^{-ad_{ij}}}{\sum_j x_j e^{-ad_{ij}}} n_i* , \tag{5.1}$$

saying that the number of people living in i and using a facility in j is dependent upon the quantity and quality of the facility in j (x_j), the physical distance between i and j, the general mobility ($1/a$) as well as the attraction exerted by the same sort of facilities in other regions. n_i* stands for all people living in i and using the facility in their own or another region.

This distribution function seems an acceptable starting point for the use of a large number of facilities. Simplifying (5.1) we write

$$\Pi_{ij} = x_j e^{-ad_{ij}} \tag{5.2}$$

and

$$\Pi_i = \sum_j x_j e^{-ad_{ij}} \tag{5.3}$$

so that (5.1) may be written as

$$n_{ij} = \frac{\Pi_{ij}}{\Pi_i} n_i* \qquad (5.4)$$

in which Π_i is the (total absolute) potential for the given facility in region i.

The demand function

A second assumption that we make regards n_i*. It seems reasonable to assume that this number depends in a positive way on the potential itself (Π_i) as well as on the average income of the i-population. Obviously, the number is also proportional to the population in i. We thus write

$$n_i* = \beta_o\ y_i^{\beta_1}\ \Pi_i^{\beta_2}\ p_i \qquad (5.5)$$

Combining (5.5) and (5.4) we obtain

$$n_{ij} = \beta_o\ y_i^{\beta_1}\ \Pi_{ij}\ \Pi_i^{\beta_2 - 1}\ p_i \qquad (5.6)$$

or

$$n_{ji} = \beta_o\ y_j^{\beta_1}\ \Pi_{ji}\ \Pi_j^{\beta_2 - 1}\ p_j \qquad (5.7)$$

from which follows

$$n*_i = \beta_o \sum_j y_j^{\beta_1}\ \Pi_{ji}\ \Pi_j^{\beta_2 - 1}\ p_j \qquad (5.8)$$

indicating the total number of users in i, in short to be called the *demand function* for the facility in question for i.

Keeping all variables except x_i constant, (5.8) may be graphically presented as figure 5.1.

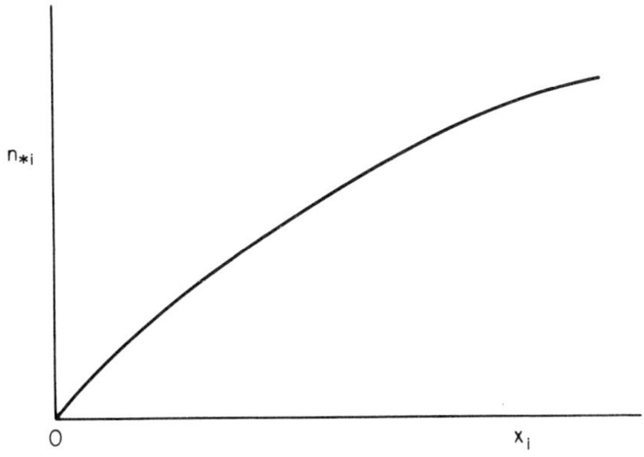

Figure 5.1 Partial relation between x_i and $n*_i$

The supply function

Now we assume that the supply of the facility depends on $n*_i$ in the following manner.

$$x_i = \gamma_o n*_i c_i^{-\gamma_1} \gamma_1 \qquad \gamma_1 \geqslant 0 \tag{5.9}$$

(5.9) is the *supply function*, in which

- c_i stands for the costs per unit of the facility (which unit is different for different facilities but usually will be a unit area).
- γ_i represents the physical planning policy in respect of region i. This quantity will be dealt with more closely later on.

In figure 5.2 the combination of supply and demand function is shown.

The equilibrium situation is reached at $(\bar{x}_i, \bar{n}*_i)$. The slope of the supply function

$$\operatorname{tg} \phi = \frac{c_i^{\gamma_1}}{\gamma_i \gamma_o} \tag{5.10}$$

and is thus an increasing function of the cost level and a decreasing function of γ_i.

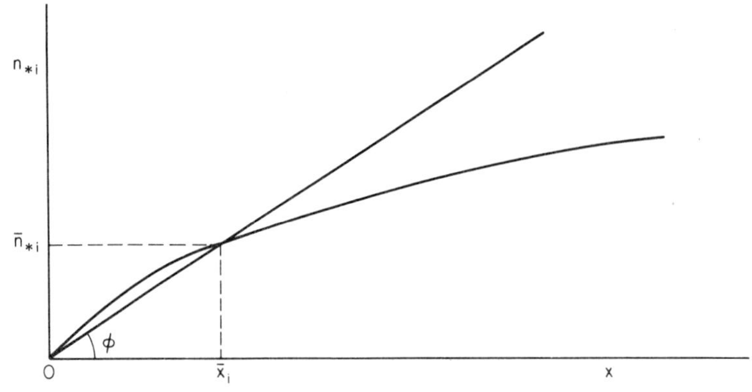

Figure 5.2 Equilibrium situation

The role fulfilled by γ_i deserves some further attention (compare also chapter 2). In principle there are four possibilities, viz.,

1.	$\gamma_i = 0$	This means an absolute prohibition to create facilities in i, a situation that may occur when i is an area of great natural beauty.
2.	$0 < \gamma_i < 1$	In this case restrictive measures are introduced, but no absolute prohibition is issued.
3.	$\gamma_i = 1$	There is no intervention with private and other plans.
4.	$\gamma_i > 1$	The creation of new facilities is stimulated by the government.

In the figure 5.3 we represent the influence of a decrease in γ_i on the equilibrium situation.

It appears that the shift of the supply function resulting from the decrease in γ_i leads to a considerable decrease in the supply of facilities and to a relatively much smaller decrease in the number of visitors. Understandingly enough, this means that restrictions imposed by the government will improve the financial position of the facilities in the 'restriction areas' or, in the opposite case, promotion and stimulation by the government will have an adverse effect on the financial position of the facilities if these are not compensated for the decrease in visitors' density.

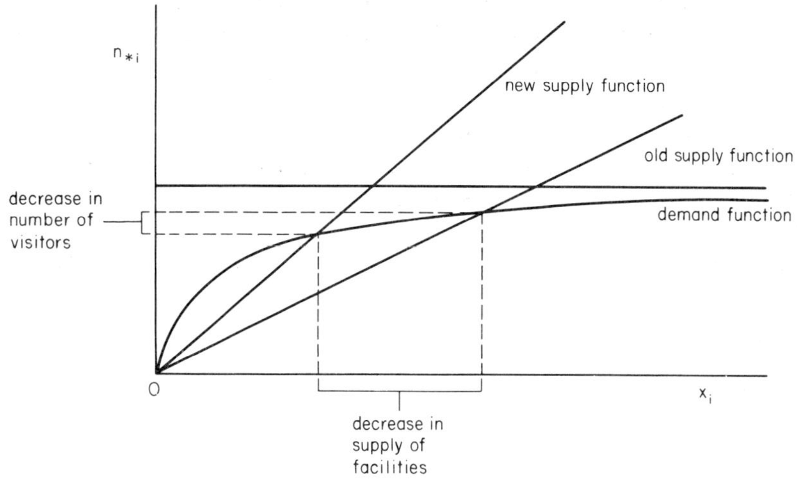

Figure 5.3 Influence of physical planning restriction

Now the approach by means of γ-values is, of course, a very general one, which covers a multitude of specific measures. In many cases the government wants to put the break on certain specified developments (e.g. construction of houses in areas of natural beauty, location of shopping centres outside towns). In that case a specific value is to be assigned to γ, for there the constraint

$$x_i \leqslant x_i^o \tag{5.11}$$

applies, in which x_i^o is the maximum permissible value of x_i. But now we can combine (5.11) with

$$x_i = x_i^* \gamma_i \tag{5.12}$$

in which x_i is actual supply and x_i^* is 'free' supply, to give

$$x_i^* \gamma_i \leqslant x_i^o$$

or

$$\gamma_i \leqslant \frac{x_i^o}{x_i^*} \tag{5.13}$$

From this equation it follows that the γ-values will have to be diminished as the 'free' value of x_i^* increases, and that means that the government will have to exert an ever stronger pressure to accomplish its objective (5.11).

The opposite is also true: promotion of facilities above the equilibrium level will entail ever higher costs, because the natural tendency to establish facilities will decrease accordingly as supply is already on a higher level.

Graphically, the above reasonings can be represented as follows:

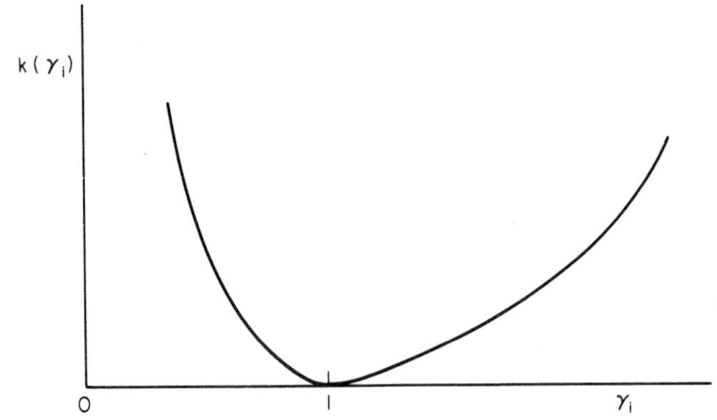

Figure 5.4 The 'costs' of physical planning intervention

At $\gamma_i = 1$ the costs are equal to zero: the government is leaving well alone. As γ_i becomes smaller, the costs are rising progressively. The pressure exerted (political pressure, too, most likely) is getting even stronger. The same holds true for values of γ_i exceeding 1. A strict and costly policy of stimulation will have to be followed, which occupies scarce means and will, therefore, meet with increasing resistance.

It is apparent that the introduction of γ-values leads to a more realistic picture than does the constraint, and we shall, therefore, keep to this approach. With that approach it is possible to make it clear why measures of spatial planning frequently prove ineffective: it is because their 'costs' exceed what is reasonable.

Related to this question is the problem of the constraints of the model. Many models contain a constraint of the form

$$\sum_k s_i^k \leqslant s_i^o - s_i^n \quad (k = \text{purposes}) \tag{5.14}$$

in words: the total use of land for all purposes together equals the total area of the region i observed less the 'unexploitable' land.

This constraint, however obvious, does not fit into the theory presented here, which assumes that the attraction factor x_i represents both quantity and quality or rather in this case area and quality. We write for purpose k as

$$x_i^k = s_i^k q_i^k \tag{5.15}$$

s_i^k = area of the facilities k

q_i^k = quality of the facilities k.

In general we may say that land-use weighted with the quality factor equals

$$x_i = \sum_k q_i^k s_i^k \tag{5.16}$$

This is one relation. The second is (5.14). They hold simultaneously and together represent the actual situation more realistically than does the first constraint by itself.

A final remark concerning q_i still is to be made. q_i represents a group of as yet undefined characteristics of facilities provided in i. Although we will be more explicit about this variable in the chapters concerning different facilities, one *diversification* index (or assortment index) needs some attention here. This variable stands for a certain quality for all the facilities in the region under consideration which is supposed to affect visiting behaviour in a certain facility over and above the quality of the facilities themselves.

Such a diversification index could rest on the following quantities:

n^k = total number of visits in the whole region paid to facility k

n = total number of visits paid to all facilities

n^{kl} = visits paid to facility k in region l

$n^l = \sum_k n^{kl}$ = total number of visits paid to region l.

Using these quantities, we can define

$$\epsilon^k = \frac{n^k}{n} \tag{5.17}$$

in which ϵ^k represents the weight of facility k in the whole region, and

$$\eta^{kl} = \frac{n^{kl}}{n^l} \tag{5.18}$$

in which η^{kl} is the weight of facility k in region l.

A diversification index can now be defined as

$$\delta_l = \left[1 - \left\{\frac{1}{2} \sum_k (\epsilon^k - \eta^{kl})^2\right\}^{\frac{1}{2}}\right] 100 \tag{5.19}$$

in which $0 \leqslant \delta_l \leqslant 100$.

If the structure of the facilities supplied in a region corresponds with the structure for the region as a whole, then

$$\epsilon^k = \eta^{kl} \ \forall k \tag{5.20}$$

and $\delta_l = 100$.

The index will be lower accordingly as the structure of the facilities supplied is farther removed from the structure represented by the supply of facilities in the region as a whole, that is to say, as it is less differentiated.

Monopolistic positions of supply and demand

The demand function (at any rate the distributive part of it) was:

$$\phi_{ij} = \frac{x_j e^{-ac_{ij}}}{\sum_j x_j e^{-ac_{ij}}} = \frac{\Pi_{ij}}{\Pi_i}$$

Now we may write

$$\frac{\partial \phi_{ij}}{\partial x_j} = \frac{1}{x_j} \left(1 - \frac{\Pi_{ij}}{\Pi_i}\right) \frac{\Pi_{ij}}{\Pi_i} \tag{5.21}$$

It is also true that

$$\frac{\partial n_{ij}}{\partial x_j} = \frac{1}{x_j} \left(1 - \frac{\Pi_{ij}}{\Pi_i}\right) \frac{\Pi_{ij}}{\Pi_i} n_i{*} \tag{5.22}$$

and, consequently, that

$$\frac{\partial \Sigma n_{ij}}{\partial x_j} = \frac{\partial n*_j}{\partial x_j} = \frac{1}{x_j} \sum_i \left(1 - \frac{\Pi_{ij}}{\Pi_i}\right) \frac{\Pi_{ij}}{\Pi_i} n_i* \tag{5.23}$$

and

$$\epsilon_j = \frac{\partial n*_j}{\partial x_j} \frac{x_j}{n*_j} = \sum_i \left(1 - \frac{\Pi_{ij}}{\Pi_i}\right) \frac{\Pi_{ij}}{\Pi_i} \frac{n_i*}{n*_j} \tag{5.24}$$

From (5.24) it follows

$$\epsilon_j = 1 - \sum_i \frac{n_{ij}}{n_i*} \cdot \frac{n_{ij}}{n*_j} \tag{5.25}$$

In this formula

$\frac{n_{ij}}{n_i*}$ represents the portion of their visits per time unit that inhabitants of i make to facilities in j

and

$\frac{n_{ij}}{n*_j}$ represents the portion of the total visits to j that is made by inhabitants of i.

Evidently

$$0 \leqslant \frac{n_{ij}}{n_i*} \leqslant 1, \quad 0 \leqslant \frac{n_{ij}}{n*_j} \leqslant 1 \text{ and } \sum_i \frac{n_{ij}}{n*_j} = 1 \text{ and } \sum_j \frac{n_{ij}}{n_i*} = 1 \tag{5.26}$$

Let

$$\frac{n_{ij}}{n_i*} = \phi_{ij} \text{ and } \frac{n_{ij}}{n*_j} = \xi_{ij}$$

then

$$\epsilon_j = 1 - \sum_i \phi_{ij} \xi_{ij} \tag{5.27}$$

in which according to (5.26)

$$0 \leqslant \sum_i \phi_{ij} \xi_{ij} \leqslant 1 \text{ and so } 0 \leqslant \epsilon_j \leqslant 1 \tag{5.28}$$

with ϕ_{ij} = 1 there is a monopoly of supply

with ξ_{ij} = 1 there is a monopoly of demand.

In the former case we speak of supply exclusivity (c.q. origin exclusivity) or selling exclusivity; in the latter of demand exclusivity (c.q. destination exclusivity) or buying exclusivity.

In fact the ξ_{ij}'s are the weights with which the expenditure quotas are weighted. Low values of the ϕ_{ij}'s mean relative unimportance of j to the customers and thus a weak selling position of region j. Customers have the possibility to choose from several alternatives; elasticity is, naturally, high, so it seems worthwhile to try and improve j's position.

High values of the ϕ_{ij}'s mean that the customer has little choice, and that elasticity is low. Improvements will have little or no effect in the face of the shops' monopolistic position and for that reason will often be omitted.

Between the ϕ's and the ξ's there is the following relation, represented for 3 regions:

$$\begin{bmatrix} n_{11} & n_{12} & n_{13} \\ n_{21} & n_{22} & n_{23} \\ n_{31} & n_{32} & n_{33} \end{bmatrix} = \begin{bmatrix} \phi_{11}n_{1*} & \phi_{12}n_{1*} & \phi_{13}n_{1*} \\ \phi_{21}n_{2*} & \phi_{22}n_{2*} & \phi_{23}n_{2*} \\ \phi_{31}n_{3*} & \phi_{32}n_{3*} & \phi_{33}n_{3*} \end{bmatrix} = \begin{bmatrix} n_{1*} & 0 & 0 \\ 0 & n_{2*} & 0 \\ 0 & 0 & n_{3*} \end{bmatrix} \begin{bmatrix} \phi_{11} & \phi_{12} & \phi_{13} \\ \phi_{21} & \phi_{22} & \phi_{23} \\ \phi_{31} & \phi_{32} & \phi_{33} \end{bmatrix} = \hat{n}_{x*}\Phi$$

But as it is also true that

$$\begin{bmatrix} n_{11} & n_{12} & n_{13} \\ n_{21} & n_{22} & n_{23} \\ n_{31} & n_{32} & n_{33} \end{bmatrix} = \begin{bmatrix} \xi_{11}n_{*1} & \xi_{12}n_{*2} & \xi_{13}n_{*3} \\ \xi_{21}n_{*1} & \xi_{22}n_{*2} & \xi_{23}n_{*3} \\ \xi_{31}n_{*1} & \xi_{32}n_{*2} & \xi_{33}n_{*3} \end{bmatrix} = \begin{bmatrix} \xi_{11} & \xi_{12} & \xi_{13} \\ \xi_{21} & \xi_{22} & \xi_{23} \\ \xi_{31} & \xi_{32} & \xi_{33} \end{bmatrix} \begin{bmatrix} n_{*1} & 0 & 0 \\ 0 & n_{*2} & 0 \\ 0 & 0 & n_{*3} \end{bmatrix} = \Xi\, \hat{n}_{*x}$$

we may say that

$$\hat{n}_{x*}\Phi = \Xi\, \hat{n}_{*x} \qquad (5.29)$$

The cases of monopoly just described represent indeed the specific application of a general theory on market concentration. That theory can be described as follows (for the sake of simplicity we will speak here in terms of purchases rather than visits).

Let

$$a = B\,1 \qquad (5.30)$$

in which: $\mathbf{a}$ = total amount of purchases by each buyer;

$\mathbf{l}$ = total amount of sales by each supplier;

$\mathbf{B}$ = square transaction matrix, a typical element of which represents the fraction of the sales by supplier j to buyer i.

Although the number of buyers need not necessarily be equal to the number of suppliers, it can be supposed without danger to the general validity that their numbers are equal. The structure of **B** determines, in fact, these numbers. In the case of shops the number of rows in the matrix equals the number of regions observed.

We could also have written

$$\mathbf{l} = \mathbf{C}\,\mathbf{a} \tag{5.31}$$

with c_{ij} representing the fraction of i's purchases bought from supplier j. Note that in this general model the 'buyer' stands for all buyers in a region, the 'seller' for all sellers in a region.

We can now write:

$$\mathbf{a} = \mathbf{B}\mathbf{l} = \mathbf{B}\mathbf{C}\mathbf{a} \tag{5.32}$$

so that

$$\mathbf{i}'\mathbf{B}\mathbf{C} = \mathbf{i}' \tag{5.33}$$

When there is a two-sided monopoly (i.e. there is only one seller and that seller has only one buyer) it is true that one element of **B** has the value of one, and all others the value of zero. In that case **B** is an elementary matrix.

$$\begin{bmatrix} a_1 \\ a_2 \\ a_3 \end{bmatrix} = \begin{bmatrix} 0 & 0 & 0 \\ 0 & 0 & 1 \\ 0 & 0 & 0 \end{bmatrix} \begin{bmatrix} l_1 \\ l_2 \\ l_3 \end{bmatrix} \tag{5.34}$$

so that $a_1 = 0$, $a_2 = l_3$ and $a_3 = 0$

while $l_1 = 0$ and $l_2 = 0$

In this situation $a = 1$ holds; (5.35)

it is comparable to the one encountered earlier, where $\phi_{ij} = 1$ and $\xi_{ij} = 1$.

With demand monopoly, the following equation holds:

$$\begin{bmatrix} a_1 \\ a_2 \\ a_3 \end{bmatrix} = \begin{bmatrix} 1 & 1 & 1 \\ 0 & 0 & 0 \\ 0 & 0 & 0 \end{bmatrix} \begin{bmatrix} l_1 \\ l_2 \\ l_3 \end{bmatrix} \tag{5.36}$$

or

$$\begin{aligned} a_1 &= l_1 + l_2 + l_3 \\ a_2 &= 0 \\ a_3 &= 0 \end{aligned} \tag{5.37}$$

which can be written as

$$\mathbf{a} = \mathbf{b'l} \text{ or } \mathbf{a} = \mathbf{i'l}$$

With supply monopoly the following equation holds:

$$\begin{bmatrix} a_1 \\ a_2 \\ a_3 \end{bmatrix} \begin{bmatrix} b_{11} & 0 & 0 \\ b_{21} & 0 & 0 \\ b_{31} & 0 & 0 \end{bmatrix} \begin{bmatrix} l_1 \\ l_2 \\ l_3 \end{bmatrix} \tag{5.38}$$

$$\text{or } \begin{aligned} a_1 &= b_{11}l_1 \\ a_2 &= b_{21}l_1 \qquad \sum_i b_{i1} = 1 \\ a_3 &= b_{31}l_1 \end{aligned} \tag{5.39}$$

In the case of (e.g. spatially) separate sub-markets with mutual monopolies it is true that

$$\mathbf{a} = \hat{\mathbf{b}}\mathbf{l}$$

Because $\hat{\mathbf{b}} = \mathbf{I}$, one may also write

$$\mathbf{a} = \mathbf{l} \tag{5.40}$$

It will be clear that such a situation may occur with facilities as an effect of distance.

A filled **B**-matrix means maximum competition, given the number of demanders and suppliers. It also implies a situation with low values of ϕ and ξ. The elasticity, ϵ_j, is very high in this case.

With m actual demanders and n actual suppliers, the number of transaction flows is equal to or smaller than nm. The sum of all elements of the transaction-coefficients matrix is equal to n. The average value of an element is, consequently, larger than $\frac{1}{n}$.

Now the so-called degree of concentration of a market may be defined as

$$\gamma' = 1 - \frac{t'}{nm} \tag{5.41}$$

in which γ' = degree of concentration (first definition)

t' = number of transaction flows (first definition)

n = number of actual suppliers

m = number of actual demanders.

An alternative, more realistic approximation would be:

$$\gamma = 1 - \frac{t}{nm} \tag{5.42}$$

in which y = degree of concentration (second definition)

t = number of transaction flows larger than a minimum value of s_o, representing, e.g., a certain proportion of the total market under consideration.

The influence of the location of the work place

In order to make the model more realistic, we will now suppose that the use of facilities is made not only from the place of living but also from the place of work. It is not really necessary to define the place of work for all users of amenities. Commercial travellers, for example, will use facilities during, after and before their work in many different places. In the following we will consider only the fixed working places.

Next to the use of facilities made from the place of work one could imagine such a use also being affected from places where one stays for some time for other reasons (e.g. places where one goes to school or for recreation), but we shall concentrate our argument on the use from places of work only.

Suppose

$$n_{ij}^{h(k)} = \frac{x_j e^{-a^h d_{ij}}}{\sum_j x_j e^{-a^h d_{ij}} + \sum_j x_j e^{-a^w d_{kj}}} n_i* = \frac{\Pi_{ij}^h}{\Pi_i^h + \Pi_k^w} n_i* \tag{5.43}$$

Here Π^{h}_{ij} is the potential of the residential place i and Π^{w}_{k} is the potential of k for people working in k. $n^{h(k)}_{ij}$ represents the visits made by an inhabitant of i working in k and using the facilities in j with his place of residence as the zone of origin. Because of the heavy constraint in time, a^{w} will usually be considerably larger than a^{h}.

The total number of visits of the group of people living in i and working in k made from their home are

$$\sum_{k} n^{h(k)}_{ij} \; p_{ik} = \Pi^{h}_{ij} \sum_{k} \left[\Pi^{h}_{i} + \Pi^{w}_{k}\right]^{-1} n_{i}{*}p_{ik} \tag{5.44}$$

in which p_{ik} represents the population dependent on those working in k.

The visits made by someone who lives in i, works in k and uses facilities in j with his place of work as zone of origin amount to

$$n^{w(k)}_{ij} = \frac{\Pi^{w}_{ij}}{\Pi^{h}_{i} + \Pi^{w}_{k}} \; n_{i}{*} \tag{5.45}$$

so that

$$\sum_{k} n^{w(k)}_{ij} \, p_{ik} = \Pi^{w}_{ij} \left[\Pi^{h}_{i} + \Pi^{w}_{k}\right]^{-1} n_{i}{*}p_{ik} \tag{5.46}$$

The total number of visits in j by inhabitants of all i's is therefore

$$n{*}_{j} = \sum_{i} \left[\Pi^{h}_{ij} + \Pi^{w}_{ij}\right] \sum_{k} \left[\Pi^{h}_{i} + \Pi^{w}_{k}\right]^{-1} n_{i}{*}p_{ik} \tag{5.47}$$

This represents the demand equation. The supply equation remains the same as in the simpler case. It will be plain that facilities far from concentration of workplaces will hardly, if at all, receive any additional visits from workers in the neighbourhood.

One or two words may be added about the quantity p_{ik}. It is, in fact, equal to $\delta_{ik} p_{i}$ when δ_{ik} represents the fraction of the active population in i that works in k. Hence, we also may write

$$p_{ik} = \delta_{ik} w_{i} \cdot \frac{p_{i}}{w_{i}} = \frac{\delta_{ik} w_{i}}{\delta_{i}} \tag{5.48}$$

δ_{i} being the participation rate of the population in i, w_{i} the total number of workers living in i and δ_{ik} the proportion of the active population from i working in k.

Obviously, the model is getting somewhat more complicated, the greatest difficulty being the estimation of the two α's, or, more accurately, of one additional α. Once the α's are known, all Π's are known too, and the result is still relatively simple.

Modal split in transportation

In the foregoing we assumed in fact one kind of trip only. In fact there is usually a choice between several modes, which, in the terms of chapter 4, gives rise to the simultaneous existence of more than one value for α, since speed, safety and money costs will be different in each case. In the following we will work out an example for two modes, r and s.

It is assumed that

$$n^r_{ij} = \frac{\Pi^r_{ij}}{\Pi^r_i + \Pi^s_i} n_i* \tag{5.49}$$

and consequently

$$n^s_{ij} = \frac{\Pi^s_{ij}}{\Pi^r_i + \Pi^s_i} n_i* \tag{5.50}$$

from which follows

$$\frac{n^r_{ij}}{n^s_{ij}} = \frac{\Pi^r_{ij}}{\Pi^s_{ij}} \tag{5.51}$$

Since

$$\sum_j n^r_{ij} = n^r_i = \frac{\Pi^r_i}{\Pi^r_i + \Pi^s_i} n_i* \tag{5.52}$$

and

$$\sum_j n^s_{ij} = n^s_i = \frac{\Pi^s_i}{\Pi^r_i + \Pi^s_i} n_i* \tag{5.53}$$

it also follows that

$$\frac{n_i^r}{n_i^s} = \frac{\Pi_i^r}{\Pi_i^s} \tag{5.54}$$

which means that the ratio of the number of trips made by mode r and mode s respectively, equals the ratio of the potentials.

Evaluation of amenity projects

Introduction

The foregoing was an exercise in models to be used in studying the use of (demand side) and provision with (supply side) amenities on the local and regional level.

It appears that these models also may be used for the evaluation of new projects consisting either of an improvement in transportation conditions or an increase in the attraction (x_j), in the form of and on consumers' surpluses based cost-benefit analysis. It will be evident, however, that this method or any other method using consumers' surpluses requires that the resistance factors all are expressed in monetary terms.

We may remind that the sum of the resistance factors per unit distance was

$$\beta_1 t + \frac{\beta_2}{V} + \beta_3 + \beta_4 \rho \tag{5.55}$$

which equals

$$\beta_1 \left(t + \frac{\beta_2}{\beta_1 v} + \frac{\beta_3}{\beta_1} + \frac{\beta_4}{\beta_1} \rho\right) \tag{5.56}$$

in which all factors between brackets are now expressed in money terms per unit of distance.

If we now write

$$\left(t + \frac{\beta_2}{\beta_1 v} + \frac{\beta_3}{\beta_1} + \frac{\beta_4}{\beta_1} \rho\right) d_{ij} = c_{ij} \tag{5.57}$$

and consider this as the weighted sum of all resistances expressed in monetary terms, we could use as an expression for the e-function to be used $e^{-\beta c_{ij}}$ in which then c_{ij} is the total sum of generalised transportation costs.

In the following we will start from an equation similar to (5.1) viz.,

$$n_{ij} = \frac{x_j e^{-\beta c_{ij}}}{\sum_j a_j e^{-\beta c_{ij}}} n_i* \tag{5.58}$$

in which a_j represents the attraction (corresponding to the quality and size of the facilities used in the preceding sections) and the other variables are already defined.

Obviously

$$\sum_j n_{ij} = n_i* \tag{5.59}$$

It will not be assumed like in some cases is done that

$$\sum_i n_{ij} = x_j \tag{5.60}$$

This point will be discussed later.

Consumers' surplus for the ij-relation

Now suppose that there is a change in c_{ij} or a comparison is made between two alternatives, each represented by a given value for c_{ij}. The situation then becomes as in figure 5.5.

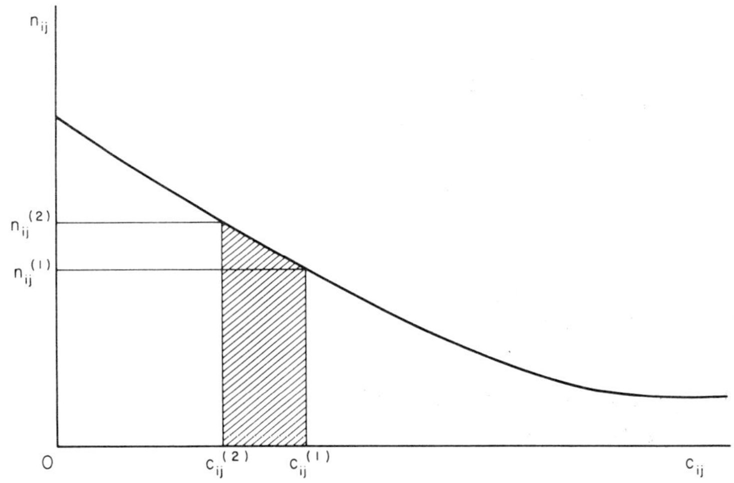

Figure 5.5 Consumers' surplus in ij- relation

As a result of the change in c_{ij} from $c_{ij}^{(1)}$ to $c_{ij}^{(2)}$ the number of trips from i to j increases from $n_{ij}^{(1)}$ to $n_{ij}^{(2)}$. The total advantage for the ij trips equals the shaded area which is the increase in consumers' surplus.

Obviously this surplus increase Δs_{ij} equals

$$\Delta s_{ij} = \int_{c_{ij}^{(2)}}^{c_{ij}^{(1)}} \frac{x_j e^{-\beta c_{ij}}}{\sum_j x_j e^{-\beta c_{ij}}} n_i dc_{ij} \qquad (5.61)$$

or

$$\Delta s_{ij} = -\frac{n_i}{\beta} \int_{c_{ij}^{(2)}}^{c_{ij}^{(1)}} \frac{dx_j e^{-\beta c_{ij}}}{\sum_j x_j e^{-\beta c_{ij}}}$$

Since $dx_j e^{-\beta c_{ij}} = d\sum_j x_j e^{-\beta c_{ij}}$

because of the fact that c_{ij} is the only variable that changes while all other c_{ik} $(k \neq j)$ remain equal, we may write

$$\Delta s_{ij} = -\frac{n_i}{\beta} \int_{c_{ij}^{(2)}}^{c_{ij}^{(1)}} \frac{d\sum_j x_j e^{-\beta c_{ij}}}{\sum_j x_j e^{-\beta c_{ij}}} \qquad (5.62)$$

Now write $\Pi_i = \sum_j x_j e^{-\beta c_{ij}}$

where Π_i represents the potential of i.

We may rewrite (5.62) as

$$\Delta s_{ij} = -\frac{n_i}{\beta} \int_{c_{ij}^{(2)}}^{c_{ij}^{(1)}} \frac{d\Pi_i}{\Pi_i} = -\frac{n_i}{\beta} \int_{c_{ij}^{(2)}}^{c_{ij}^{(1)}} d\ln\Pi_i \tag{5.63}$$

so that

$$\Delta s_{ij} = -\frac{n_i}{\beta} \left\{ \ln \Pi_i^{(1)} - \ln\Pi_i^{(2)} \right\} = -\frac{n_i}{\beta} \ln \frac{\Pi_i^{(1)}}{\Pi_i^{(2)}}$$

$$\boxed{\Delta s_{ij} = -\frac{n_i}{\beta} \ln \frac{\Pi_i^{(1)}}{\Pi_i^{(2)}}} \tag{5.64}$$

It may be noted that because of the improvement in traffic conditions ($c_{ij}^{(1)} > c_{ij}^{(2)}$), $\Pi_i^{(2)} > \Pi_i^{(1)}$. The potential of i has increased in size. This implies that

$$\ln \frac{\Pi_i^{(1)}}{\Pi_i^{(2)}} < 0 \text{ and thus } \Delta s_{ij} > 0.$$

Consumers' surplus for the ik-relations

The change in the ik-demand relation is presented in figure 5.6.

The demand curve has shifted downwards as a result of the decrease in c_{ij}. Because of this decrease a number of ik trips equalling Δn_{ij} has shifted towards ij-trips. The advantages sprouting from this shift have already been counted in the consumers' surplus for the ij-relation. The only change in surplus left is thus for the remaining ik-trips ($n_{ik}^{(2)}$). In fact this change is represented by the difference between the two shaded areas in the graph.

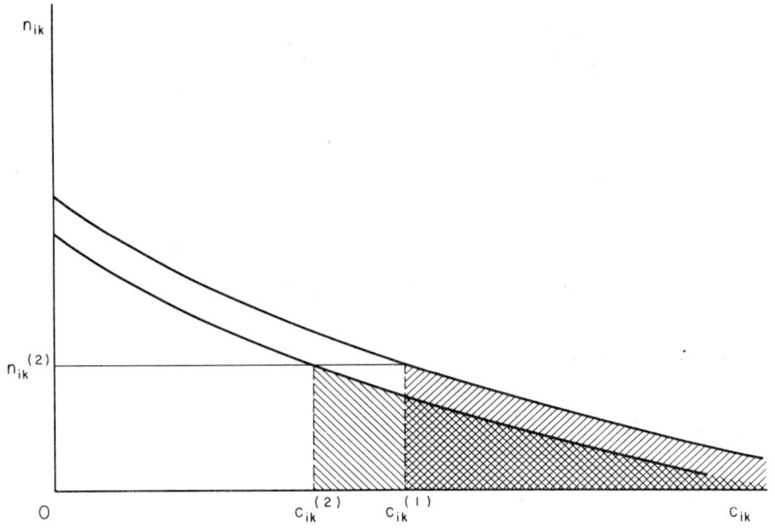

Figure 5.6 Consumers' surpluses in the ik-relation

This difference can be calculated as follows.

$$s_{ik} = n_i \int_{c_{ik}}^{\infty} \frac{x_k e^{-\beta c_{ik}}}{\sum_k x_k e^{-\beta c_{ik}}} dc_{ik} \tag{5.65}$$

$$= -\frac{n_i}{\beta} \int_{c_{ik}}^{\infty} \frac{d\sum_k x_k e^{-\beta c_{ik}}}{\sum_k x_k e^{-\beta c_{ik}}} = -\frac{n_i}{\beta} \left[\ln \Pi_i \right]_{c_{ik}}^{\infty} \tag{5.66}$$

Now write $\lim_{c_{ik} \to \infty} \Pi_i = \lim_{c_{ik} \to \infty} \sum_k x_k e^{-\beta c_{ik}} = \Pi_i - x_k e^{-\beta c_{ik}} = \Pi_i^{\not{k}}$ (5.67)

Thus $s_{ik} = -\frac{n_i}{\beta} \ln \frac{\Pi_i^{k\!\!\!/}}{\Pi_i}$ (5.68)

Now the decrease in c_{ij} may be translated for $n_{ik}^{(2)}$ into an increase in the price c_{ik}. Indicate these (simulated) prices by $c_{ik}(I)$ and $c_{ik}(II)$. Then the change in consumers' surplus can be represented by

$$\Delta s_{ik} = \frac{n_i}{\beta} \ln \frac{\Pi_i^{k\!\!\!/}(I)}{\Pi_i(I)} \cdot \frac{\Pi_i(II)}{\Pi_i^{k\!\!\!/}(II)} \tag{5.69}$$

which should hold under the condition that $n_{ik}(I) = n_{ik}(II)$.

Now write

$$\Delta s_{ik} = -\frac{n_i}{\beta} \ln \frac{\dfrac{\Pi_i^{k\!\!\!/}(I)}{\Pi_i(I)}}{\dfrac{\Pi_i^{k\!\!\!/}(II)}{\Pi_i(II)}} = -\frac{n_i}{\beta} \ln \frac{\dfrac{\Pi_i(I) - x_k e^{-\beta c_{ik}(I)}}{\Pi_i(I)}}{\dfrac{\Pi_i(II) - x_k e^{-\beta c_{ik}(II)}}{\Pi_i(II)}}$$

$$= -\frac{n_i}{\beta} \ln \frac{1 - \dfrac{x_k e^{-\beta c_{ik}(I)}}{\Pi_i(I)}}{1 - \dfrac{x_k e^{-\beta c_{ik}(II)}}{\Pi_i(II)}}$$

$$= -\frac{n_i}{\beta} \ln \frac{1 - \dfrac{n_{ik}(I)}{n}}{1 - \dfrac{n_{ik}(II)}{n}} = 0 \tag{5.70}$$

since $n_{ik}(I) = n_{ik}(II)$

We may thus say that the total change in consumers' surplus equals the surplus for the ij-relation.

An approximation

The question arises how the expression (5.64) can be simplified.

$$\text{Since} \quad \frac{\Pi_i^{(1)}}{\Pi_i^{(2)}} = e^{-\beta\{c_{ij}^{(1)} - c_{ij}^{(2)}\}} \quad \frac{n_{ij}^{(2)}}{n_{ij}^{(1)}} \tag{5.71}$$

we may write

$$\Delta s_{ij} = -\frac{n_i}{\beta} \ln \{1 + \frac{\Delta n_{ij}}{n_{ij}^{(1)}}\} + n_i \{c_{ij}^{(1)} - c_{ij}^{(2)}\}$$

$$\widetilde{\Delta s}_{ij} = + \frac{n_i}{\beta} \frac{\Delta n_{ij}}{n_{ij}^{(1)}} + n_i \{c_{ij}^{(1)} - c_{ij}^{(2)}\} \tag{5.72}$$

Now write

$$\frac{dn_{ij}}{dc_{ij}} = -\beta \frac{n_{ij}}{n_i} \{1 - \frac{n_{ij}}{n_i}\} n_i \tag{5.73}$$

and

$$\frac{dn_{ij}}{n_{ij}} = -\frac{\beta dc_{ij}}{n_i} \{1 - \frac{n_{ij}}{n_i}\} n_i$$

$$\text{then} \qquad \frac{dn_{ij}}{n_{ij}} \frac{n_i}{\beta} = -dc_{ij} \{n_i - n_{ij}\} \tag{5.74}$$

We now approximate the value of

$\frac{\Delta n_{ij}}{n_{ij}} \frac{n_i}{\beta}$ by the average of the two values of (5.74) for $n_{ij}^{(1)}$ and $n_{ij}^{(2)}$.

We may then write

$$\frac{\Delta n_{ij}}{n_{ij}^{(1)}} \cdot \frac{n_i}{\beta} = -\Delta c_{ij} \{n_i - \frac{n_{ij}^{(1)} + n_{ij}^{(2)}}{2}\} \tag{5.75}$$

Substituting this value into (5.72) we obtain:

$$\Delta\tilde{s}_{ij} = -n_i \Delta c_{ij} + \frac{n_{ij}^{(1)} + n_{ij}^{(2)}}{2} \Delta c_{ij} + n_i \Delta c_{ij}$$

$$\boxed{\Delta\tilde{s}_{ij} = \frac{n_{ij}^{(1)} + n_{ij}^{(2)}}{2} \Delta c_{ij}} \qquad (5.76)$$

which means that the approximate value of the change in the consumers' surplus on the ij-relation equals the product of the cost difference and the average number of trips in both situations.

We write

$$\boxed{\Delta\tilde{s}_{ij} = \overline{n_{ij}} \Delta c_{ij}} \qquad (5.77)$$

A simple example

We take an example in which

$$c_{ij}^{(1)} = 10$$

$$c_{ij}^{(2)} = 9$$

$$n_{ij} = 100$$

$$n_i = 10{,}000$$

$$\beta = 0.1$$

Substituting these values into (5.72) we obtain

$$\Delta\tilde{s}_{ij} = -\frac{\Delta n_{ij}}{100} \cdot \frac{10{,}000}{0.1} + 10{,}000\,(10\text{-}9)$$

$$= -\ 1{,}000\,\Delta n_{ij} + 10{,}000$$

A minimum estimate for $\Delta\tilde{s}_{ij}$ is $n_{ij}\ \{c_{ij}^{(1)} - c_{ij}^{(2)}\}$
This value is reached if $\Delta n_{ij} = 0$. In this case only the trips in situation 1 profit from the decrease in c_{ij}, but no new trips are created.

We thus find

$$\Delta \tilde{s}_{ij} = n_{ij} \{c_{ij}^{(1)} - c_{ij}^{(2)}\} = 100\,(10\text{-}9) = 100.$$

So $100 = -\,1{,}000\,\Delta\, n_{ij} + 10{,}000$

then $\qquad -\,9{,}900 = -\,1{,}000\,\Delta\, n_{ij}$

or $\Delta n_{ij} = 9.9.$

As the elasticity of n_{ij} with respect to c_{ij} roughly equals $-\beta c_{ij}$, the value of the elasticity in this case is approximately equal to -1. The value found on the basis of a decrease in c_{ij} of 10 per cent is $-\,0.99$ which corresponds very closely to the expectation.

Consumers' surplus in the case of a modal split

The consumers' surpluses are calculated in exactly the same way as for one mode. In fact there is no difference in assuming either that there are two modes to reach j from i, or that there are two j's to be reached with the same mode. Both assumptions are formally identical.

Write (like we did in the section 'Modal split in transportation', p.89)

$$n_{ij}^{r} = \frac{\Pi_{ij}^{r}}{\Pi_i^{r} + \Pi_i^{s}}\; n_i \qquad (5.78)$$

then

$$\Delta s_{ij}^{r} = n_i \int_{c_{ij}^{(2)}}^{c_{ij}^{(1)}} \frac{\Pi_{ij}^{r}}{\Pi_i^{r} + \Pi_i^{s}}\, dc_{ij}^{r}$$

$$= -\frac{n_i}{\beta} \int_{c_{ij}^{(2)}}^{c_{ij}^{(1)}} d \ln (\Pi_i^{r} + \Pi_i^{s})$$

$$= -\frac{n_i}{\beta} \ln \frac{(\Pi_i^{r} + \Pi_i^{s})\ (1)}{(\Pi_i^{r} + \Pi_i^{s})\ (2)} \qquad (5.79)$$

which is formally identical to (5.64).

The integral for s_{ik}^r is

$$s_{ik}^{r} = n_i \int_{c_{ik}^r}^{\infty} \frac{x_k e^{-\beta c_{ik}^r}}{\sum_k x_k e^{-\beta c_{ik}^r} + \sum_k x_k e^{-\beta c_{ik}^s}} \, dc_{ik}^r \tag{5.80}$$

$$= -\frac{n_i}{\beta} \int_{c_{ik}^r}^{\infty} \frac{d\sum_k x_k e^{-\beta c_{ik}^r}}{\sum_k x_k e^{-\beta c_{ik}^r} + \sum_k x_k e^{-\beta c_{ik}^s}}$$

$$= -\frac{n_i}{\beta} \int_{c_{ik}^r}^{\infty} \frac{d\{ \sum_k x_k e^{-\beta c_{ik}^r} + \sum_k x_k e^{-\beta c_{ik}^s} \}}{\sum_k x_k e^{-\beta c_{ik}^r} + \sum_k x_k e^{-\beta c_{ik}^s}}$$

$$= -\frac{n_i}{\beta} \int_{c_{ik}^r}^{\infty} d \ln (\Pi_i^r + \Pi_i^s) \tag{5.81}$$

So

$$\Delta s_{ik}^{r} = \frac{n_i}{\beta} \ln \frac{(\Pi_i^r + \Pi_i^s)^{k^r} \text{(I)}}{(\Pi_i^r + \Pi_i^s)^{k^r} \text{(II)}} \cdot \frac{(\Pi_i^r + \Pi_i^s) \text{ (II)}}{(\Pi_i^r + \Pi_i^s) \text{ (I)}} \tag{5.82}$$

which is formally identical to (5.69).

Result for the modal split case

Since all approximations are identical to the ones already carried out, we may conclude that a close approximation of the change in consumers' surpluses is represented by

$$\Delta s^r = \sum_i \sum_j \bar{n}_{ij}^r \{c_{ij}^r(1) - c_{ij}^r(2)\}$$

$$\Delta s^s = \sum_i \sum_j \bar{n}_{ij}^s \{c_{ij}^s(1) - c_{ij}^s(2)\}$$

or

$$\boxed{\Delta s = \Delta s_i^r + \Delta s_i^s = \sum_i \sum_j [\bar{n}_{ij}^r \{c_{ij}^r(1) - c_{ij}^r(2)\} + \bar{n}_{ij}^s \{c_{ij}^s(1) - c_{ij}^s(2)\}]}$$

(5.83)

This means that the calculation of the benefits can proceed as follows:

1. Given

   (a) the generalised costs matrices $\mathbf{C}_1^k = [c_{ij}^k(1)]$

   and $\mathbf{C}_2^k = [c_{ij}^k(2)]$

   for each mode k in situations 1 and 2;

   (b) the trip matrix $\underset{\sim}{N}_1^k = [t_{ij}^k(1)]$

   $\underset{\sim}{N}_2^k = [t_{ij}^k(2)]$

2. The benefits may be estimated at

   $$B = \tfrac{1}{2} \sum_k [i' \{\underset{\sim}{N}_1^k + \underset{\sim}{N}_2^k\} \circledS \{C_1^k - C_2^k\}] \; i \qquad (5.84)$$

   in which $i = \begin{bmatrix} 1 \\ 1 \\ 1 \end{bmatrix}$ and Ⓢ is the Schur-product in the sense that

   e.g. $[c_{ij}^k(1)] \;\; s \;\; [n_{ij}^k(1)] = [c_{ij}^k(1)\, n_{ij}^k(1)]$

Improved availability of facilities

The preceding chapters related to an improvement of transportation facilities. The second possibility for improvement is the increased availability of facilities.

We write:

$$C_i = n_i * \int_o^\infty \frac{x_j e^{-\beta c_{ij}}}{\sum_j x_j e^{-\beta c_{ij}}} dc_{ij} \tag{5.85}$$

for the consumers' surplus that people living in i derive from their trips to j.

Write

$$z_i = x_j e^{-\beta c_{ij}}$$

and

$$\sum_k x_k e^{-\beta c_{ij}} = z_i + c_i = \sum_j x_j e^{-\beta c_{ij}}$$

in which, of course

$$c_i = \sum_{k \neq j} x_k e^{-\beta c_{ik}}$$

Then

$$dz_i = - \beta z_i dc_{ij} \tag{5.86}$$

so may be rewritten as

$$C_i = \frac{n_i *}{\beta} \int_o^{a_j} \frac{dz_i}{z_i + c_i} = \frac{n_i *}{\beta} \ln (1 + \frac{x_j}{c_i}) \tag{5.87}$$

The increase in C_i resulting from an increase in a_j with Δa_j can now be calculated as follows

$$\Delta c_i = \frac{n_i*}{\beta} \ln\left(1 + \frac{\Delta x_j}{c_i + x_j}\right) \cong \frac{n_i*}{\beta} \frac{\Delta x_j}{c_i + x_j} \tag{5.88}$$

The total increase for all relations with j is thus

$$\Delta C = \frac{n_i*}{\beta} \Delta x_j \sum_i \frac{1}{c_i + x_j} \tag{5.89}$$

The other relations do not give rise to any change in consumers' surplus.

The foregoing types of analysis will enable us to evaluate the effects of transportation improvements or increased availability of facilities. It is obvious that for this purpose data on the flows of visitors (or money flows) should be available.

Some special features of a shopping model

Introduction

Although basically in every amenity model there will be simultaneously money flows and flows of users or visitors, in a shopping model this feature is so outspoken that it deserves special attention in a separate section. In the following we will point at the consequences of the introduction of money flows and visitor flows simultaneously and discuss some general features of a shopping model in so far as these have not been treated fully in the preceding paragraphs.

As a starting point, four quantities, all related to one simple kind of shops (say, shops selling foodstuffs or household durables), will be defined. To keep the notation as simple as possible, no subscript denoting the kind of shop will be used. The four quantities are:

f_{ij} = purchases made per unit of time by inhabitants of region i in region j per head of the population in i (purchase amounts) (i, j = 1, 2, . . . , n);

q_j = the quality index of the shopping provision (in the category concerned) in j at a certain moment. q_j includes the diversification index which in this case might be called assortment index;

s_j = the floor area of shopping provision in j at a certain moment;

c_{ij} = generalised transportation costs.

For the sake of simplicity we will use also

$$x_j = q_j s_j$$

representing the attraction resulting from size and quality of the shopping facilities in the branch under consideration.

The demand function

We write similar as before:

$$f_{ij} = \frac{x_j e^{-\beta c_{ij}}}{\sum_j x_j e^{-\beta c_{ij}}} f_{i*} = \frac{\Pi_{ij}}{\Pi_i} f_{i*} \tag{5.90}$$

in which f_{i*} is the sum of all expenditures of inhabitants of i on the goods sold in the branch studied.

Note that the quantity Π_{ij}/Π_i approaches unity as β approaches infinity. In that case (daily goods) the consumer buys (almost) exclusively in his own region.

We assume furthermore that the average annual expenses made by consumers are proportional with their average annual disposable income, so that

$$f_{i*} = \gamma y_i \tag{5.91}$$

where y_i represents the disposable annual income per head in i, and γ the fraction of income spent on the kind of goods sold in the category of shops involved. Other, perhaps even more realistic expressions for this function could be imagined, such as

$$f_{i*} = \gamma_o y_i^{\gamma_1}, \text{ with } \gamma_1 > 1 \text{ for luxuries and}$$
$$\gamma_1 < 1 \text{ for necessities}$$

$$f_{i*} = \gamma_o + \gamma_1 y_i, \text{ with } \gamma_o < 0 \text{ for luxuries and}$$
$$\gamma_o > 0 \text{ for necessities.}$$

Application of these formulas would not essentially change our shopping model, in which income is considered exogenous. So, we may conveniently stick to (5.91), and go on to write:

$$F_{ij} = p_i f_{ij} = p_i \frac{\Pi_{ij}}{\Pi_i} f_{i*} = \frac{\Pi_{ij}}{\Pi_i} F_{i*} \qquad (5.92)$$

in which F_{ij} is the total amount spent in j by inhabitants of i, and p_i the volume of population in i.

Of the total amount spent in j it can be said that

$$F_{*j} = \sum_i F_{ij} = \sum_i p_i \frac{\Pi_{ij}}{\Pi_i} f_{i*} = \gamma \sum_i \frac{\Pi_{ij}}{\Pi_i} Y_i \qquad (5.93)$$

where $Y_i = p_i y_i$, i.e. the total disposable annual income available to inhabitants of i.

(5.93) is called the *demand function.*

It represents the demand for shopping services in the category involved, expressed in the amount spent in j as a function of the facilities available there and elsewhere.

The supply function

Now we assume that the total supply of shops in the category under observation in j depends on F_{*j} in the following manner:

$$x_j = \beta_o \left(\frac{F_{*j}}{c_j} \right)^{\beta_1} \Gamma_j \qquad (5.94)$$

in which presumably $\beta \approx 1$

and $\Gamma_j \geqslant 0$

(5.94) is the *supply function*, in which

c_j stands for the cost index per unit of shopping area in j. In this index are included the costs of land, buildings, staff, etc. On an average it is true of all regions that

$\bar{c} = 1.$

Γ_j represents the physical-planning policy in respect of region j.

If $\beta_1 = 1$, then

$$x_j = \beta_o \frac{F*_j}{c_j} \Gamma_j,$$

or

$$\frac{F*_j}{s_j} = \frac{q_j c_j}{\Gamma_j \beta_o} \quad (5.95)$$

which means that the turnover per unit of floor area in j is proportional to quality and cost level, and inversely proportional to Γ_j. Stimulation of the shopping function ($\Gamma_j > 1$) will, therefore, cause the turnover per area unit to decrease; discouragement ($\Gamma_j < 1$) causes the turnover to increase. The equilibrium position can be derived easily from (5.94) and (5.93).

Frequency of shopping trips

The average expenditure was treated in the previous pages as the average amount spent per period. One might, of course, also be interested in the average amount spent per visit as well as in the total number of visits made. The former quantity can be defined as

$$\bar{f}_{ij} = \frac{f_{ij} p_i}{v_{ij}} \quad (5.96)$$

in which v_{ij} is the number of trips inhabitants of i make per period to the kind of shops we are studying in j, that is to say, the visiting frequency.

Now let us assume

$$V_{ij} = \frac{x_j e^{-\delta c_{ij}}}{\sum_j x_j e^{-\delta c_{ij}}} V_i* = \frac{N_{ij}}{N_i} V_i* \quad (5.97)$$

where V_i* represents the number of trips inhabitants of i make for the purpose in question.

We assume further that

$$V_i* = a_o p_i (N_i)^{a_1} y_i^{a_2} \quad (5.98)$$

So, the number of trips depends on the attraction potential for trips (N_i) and the income, and is proportional to the population in i. Now it is true that

$$V_{ij} = a_o \frac{N_{ij}}{N_i} p_n N_i^{a_1} y_i^{a_2}$$

$$= a_o p_i N_{ij} N_i^{a_1 - 1} y_i^{a_2} \qquad (5.99)$$

The average amount spent per trip follows from dividing F_{ij} by V_{ij} :

$$\bar{f}_{ij} = \frac{F_{ij}}{V_{ij}} = \frac{p_i f_{ij}}{a_o p_i N_{ij} N_i^{a_1-1} y_i^{a_2}} = \frac{\gamma}{a_o} \frac{\Pi_{ij}}{\Pi_i} \frac{N_i}{N_{ij}} N_i^{-a_1} y_i^{1-a_2} \qquad (5.100)$$

if $a = \delta$, then $\Pi_{ij} = N_{ij}$

and $\Pi_i = N_i$

so that

$$\bar{f}_{ij} = \frac{\gamma}{a_o} N_i^{-a_1} y_i^{1-a_2} \qquad (5.101)$$

Evidently, the average amount spent decreases with a rising shopping potential. Because $a_2 < 1$, income exerts a positive influence on the visiting frequency.

Concluding remarks. Migration

Although it would be possible to present models for other activities (schools, recreation, sports facilities, etc.) it would add so little to our knowledge compared to what we already presented in the general model that it hardly pays to write it all down or to read it. The basic principles are identical.

What seems worthwhile to indicate however, is that, with the foregoing we have treated a kind of model that could provide us with knowledge about a number of operational variables which we have indicated as potentials that describe the attractivity of a region for residential purposes. Since differences between degrees of attractiveness of regions could be important explaining factors for migration, at least with the foregoing we have made one step in the construction of a migration function. In chapter 3 we introduced already such a function and it will be our task in the next chapters to complement this function with the other elements indicated in chapter 3, viz., with the elements related to the job opportunities.

References

Applebaum, William, 'How to Measure the Value of a Trading Area', *Chain Store Age*, November 1940.

Berry, Brian, J.L., *Geography of Market Centres and Retail Distribution*, Prentice Hall, N.J., 1967.

Berry, Brian, J.L., Gardiner, H. and Tennant, Robert J., 'Retail Location and Consumer Behaviour', *Papers and Proceedings of the Regional Science Association*, vol.9, 1962, pp 65-106.

Bucklin, Louis P., *Consumer Shopping Patterns in an Urban Region*, Berkeley School of Business – University of California, 1966.

The Economist, 'Shopping Goes Out of Town', London 15-21 April 1972, p.60 ff.

Huff, David L., 'A Probabilistic Analysis of Shopping Trade Centre Areas', *Land Economics*, vol.XXXIX, no.1, 1963.

Huff, David L, en Larry Blue, 'A programmed Solution for Estimating Retail Sales Potentials', Lawrence Kamai: Centre for Regional Studies, 1966.

Klaassen, Leo H., *Urban Planning and its impact on the Quality of Life in European Cities*, NEI, Rotterdam, 1972.

Lakshmanan, T.R. and Ilamen, W.G. 'A Retail Market Potential Model', *Journal of the American Institute of Planners*, vol.31, September 1975.

Luce, R. Duncan, *Individual Choice Behavior*, Wiley and Sons, New York, 1958.

National Economic Development Office, *Urban models in shopping studies*, London, 1970.

Reilly, William, J., Ph.D., *The Law of Retail Gravitation*, second edition, New York, 1953.

Ruitenberg, A.A., *De Dordtse Binnenstad*, NEI, Rotterdam, 1973.

Voorhees, Alan M., Sharpe, Gordon B. and Stegmaier, J.T., *Shopping Habits and Travel Patterns*, Urban Land Institute, Technical Bulletin no.24.

Wagner, Louis C., *Economic Relationship of Parking to Business in Seattle Metropolitan Area*, Highway Research Board, Washington D.C., 1953.

6. Interregional attraction models (SPAMOI)

Introduction

In his 'Methods of Selecting Industries for Depressed Areas' (1967), Klaassen proposes a method for selecting activities that might stimulate economic growth in such areas. A fundamental assumption underlying that method is that the scale of the activities of an industry in a given region will be determined by the demand for its products in that region as well as the local supply of its requirements. [1] The expression symbolising the relation between the quantities mentioned is called by Klaassen the *attraction equation* of the industry concerned in the given region.

After Klaassen, Van Wickeren (1971), Paelinck (1973), and Deacon (1975), among others, have formulated attraction equations. In a later section of this chapter it will be described how Klaassen in the study cited above derives attraction equations for a closed regional system. Van Wickeren (1971) introduced the concepts of *attraction theory, specific attraction model* and *general attraction model*. These concepts will then be discussed, and the specific and general attraction models examined to find out if they are *structural models*.

An industry located in a certain region can, in principle, sell its products both inside and outside that region. The flow of goods and services sold outside the region can be divided into, on the one hand, a flow to regions that form one *regional system* with the region of location, and on the other hand, a flow of goods and services sold outside this system. As soon as one succeeds in explaining both the sales to each of the regions inside the regional system and the total sales to regions outside that system, the scale of the activities of an industry in a given region will be determined as well, on the condition that the sum of sales equals the production. This condition can also be defined as follows: 'the volume of any stocks of the industry involved should remain constant', a condition that will nearly always be fulfilled in the long run.

Four models will be developed, to which the explanation of the sales to each of the regions in a regional system will be central. For each of the four models the assumption holds that no industry sells goods or services to, or imports products from, regions outside the regional system under observation. This assumption has been made for the sake of simplicity; further research will among other things be directed towards formulating models to which the assumption does not apply. This chapter will close with a number of conclusions.

The attraction equations derived by Klaassen for a closed regional system

The selection method cited in the introduction was based upon the idea that transportation costs are not decisive for the locational decisions of entrepreneurs but rather *communication costs* in the widest sense of the word. [2]

The first step towards developing an attraction equation is, therefore, the definition of the communication costs for a given industry. Generally speaking, these costs consist of the communication costs involved in selling the industry's production on output markets and acquiring the inputs it needs on input markets. Klaassen assumes that such input and output markets are confined within a regional system, [3] comprising, at a given regional division, N regions. In relation to industry k ($k = 1, \ldots, M$), he then distinguishes two zones. The first zone is the area within which both the communication costs per unit of product to be sold and the communication costs involved in attracting a unit of input – for each input factor needed – tend towards zero; this zone is called the *relevant region.* [4] The second zone is the area lying outside the relevant region but within the regional system. An industry k, located within region i as well as within its relevant region, might export part of its production to the latter zone and import part of the inputs required for its production from this zone.

In most cases no figures are available with respect to the communication costs per unit of output to be sold or per unit of input to be obtained, for such communications have up until now hardly been established; consequently it is not possible directly to determine an industry's relevant region. Klaassen has developed a procedure, however, which enabled him to determine the relevant region in an indirect way. [5]

Let us now introduce the assumption that in what follows the regional system will be divided in such a way that the region i in which industry k is located, is at the same time its relevant region. In that case, the second zone referred to above will henceforth comprise all the regions of the spatial system except i. It follows from this assumption that within region i of the regional system the communication costs, both per unit of product to be sold as per unit of input to be acquired, tend towards zero.

Klaassen defines the communication costs for industry k in region i as:

$$c_i^k \triangleq \tau_{iw}^k (q_i^k - d_{*_i}^k) + \sum_{l=1}^{M} \tau_{wi}^{lk} \; rr_{wi}^{lk} \quad (k = 1, \ldots, M) \qquad (6.1)$$

The symbols in equation (6.1) have the following meaning:

- c_i^k the communication costs for industry k in region i;
- τ_{iw}^k the communication costs involved in the supply of one unit by industry k in region i to region w, for the account of industry k in region i, region w representing the second zone;
- q_i^k the production of industry k in region i;
- $d_{*_i}^k$ the demand in region i for products of industry k located within the regional system;
- τ_{wi}^{lk} the communication costs involved in the supply of one unit by industry l in region w to industry k in region i, for the account of industry k in region i; [6]
- rr_{wi}^{lk} the supply by industry l in region w to industry k in region i.

The export of industry k in region i to the other regions within the regional system is represented by $q_i^k - d_{*_i}^k$. Because the communication costs per unit figure as constants in (6.1), it is assumed that τ_{iw}^k and τ_{wi}^{lk} are independent of both the physical distance between regions i and w and the social distance between industries k and l. Furthermore it appears from this equation that τ_{ii}^k and τ_{ii}^{lk} have been put at zero.

In the second step towards formulating the attraction equation for industry k in region i, rr_{wi}^{lk} is elaborated by means of the identity:

$$rr_{*_i}^{lk} \equiv rr_{ii}^{lk} + rr_{wi}^{lk} \qquad (k, l = 1, \ldots, M) \qquad (6.2)$$

in which

$rr_{*_i}^{lk}$ represents the supply by industry l established within the regional system to industry k in region i;

and

rr_{ii}^{lk} represents the supply by industry l in region i to industry k in region i.

With the help of the following definition of the technical and the allocation coefficient, respectively:

$$\beta^{lk}_{**} \triangleq \frac{\sum_i rr^{lk}_{*_i}}{\sum_i q^k_i}; \quad a^{lk}_{**} \triangleq \frac{\sum_i rr^{lk}_{*_i}}{\sum_i q^l_i} \qquad (k, l = 1, \ldots, M) \tag{6.3}$$

Klaassen writes $rr^{lk}_{*_i}$ as

$$rr^{lk}_{*_i} \triangleq \beta^{lk}_{**}\, q^k_i \qquad (k, l = 1, \ldots, M) \tag{6.4}$$

and rr^{lk}_{ii} as

$$rr^{lk}_{ii} \triangleq a^{lk}_{**}\, q^l_i \qquad (k, l = 1, \ldots, M) \tag{6.5}$$

so that

$$rr^{lk}_{wi} = \beta^{lk}_{**}\, q^k_i - a^{lk}_{**}\, q^l_i \qquad (k, l = 1, \ldots, M) \tag{6.6}$$

The third step comprises closer observation of c^k_i, the communication costs for industry k in region i. Klaassen poses that these costs can also be looked upon as the value of the services rendered by the transportation and communication industries in region i to industry k in the same region; evidently Klaassen assumed that industry k in region i does not import any transportation and communication services from the other regions of the regional system. Thus, the communication costs for industry k in region i are written as the product of the allocation coefficient a^{tk}_{**} and the production of the transportation and communication industry in region i, q^t_i. In symbols:

$$c^k_i \triangleq a^{tk}_{**}\, q^t_i \qquad (k = 1, \ldots, M) \tag{6.7}$$

In the fourth step the attraction equations are derived from the relations (6.1), (6.3), (6.6) and (6.7). [7]

$$c^k_i = (\tau^k_{iw} + \sum_{l \neq t} \tau^{lk}_{wi} \beta^{lk}_{**}) q^k_i - \tau^k_{iw}\, d^k_{*_i} - \sum_{l \neq t} \tau^{lk}_{wi}\, a^{lk}_{**}\, q^l_i \qquad (k = 1, \ldots, M) \tag{6.8}$$

or

$$q^k_i = \frac{1}{\tau^k_{1w} + \sum_{l \neq t} \tau^{lk}_{w1}\, \beta^{lk}_{**}} (\tau^k_{1w}\, d^k_{*_i} + \sum_{l \neq t} \tau^{lk}_{w1}\, a^{lk}_{**} q^l_i + c^k_i) \tag{6.9}$$

$$(k = 1, \ldots, M)$$

in which use is made of the assumption that $\tau^k_{iw} = \tau^k_{1w}$ and $\tau^{lk}_{wi} = \tau^{lk}_{w1}$. We put

$$\lambda(k) \triangleq \frac{\tau^k_{1w}}{\tau^k_{1w} + \sum_{l \neq t} \tau^{lk}_{w1} \beta^{lk}_{**}} \; ; \; \lambda(l, k) \triangleq \frac{\tau^{lk}_{w1} \beta^{lk}_{**}}{\tau^k_{1w} + \sum_{l \neq t} \tau^{lk}_{w1} \beta^{lk}_{**}} \tag{6.10}$$

$$(k, l = 1, \ldots, M; l \neq t)$$

and

$$\lambda(l,k) \triangleq \frac{\beta^{tk}_{**}}{\tau^k_{1w} + \sum_{l \neq t} \tau^{lk}_{w1} \beta^{lk}_{**}} \qquad (k = 1, \ldots, M; l = t) \tag{6.11}$$

then from (6.7) and (6.9) through (6.11) an attraction equation for industry k in region i follows, viz.:

$$q^k_i = \lambda(k)\, d^k_{*_i} + \sum_l \lambda(l,k) \frac{a^{lk}_{**}}{\beta^{lk}_{**}} q^l_i \qquad (k = 1, \ldots, M) \tag{6.12}$$

This attraction equation represents the relation between, on the one hand, the output of industry k in region i, and on the other, the demand in region i for the products of that industry and the local production of the inputs that industry k in region i needs to produce its output. The relation tested by Klaassen is not that of (6.12), but the following one. [8]

$$\frac{q^k_i}{q^k_*} = \lambda(k) \frac{d^k_{*_i}}{d^k_{**}} + \sum_l \lambda(l,k) \frac{q^l_i}{q^l_*} \qquad (k = 1, \ldots, M) \tag{6.13}$$

in which

$$q^k_* \triangleq \sum_i q^k_i \quad \text{and} \quad d^k_{**} \triangleq \sum_i d^k_{*_i} \qquad (k = 1, \ldots, M) \tag{6.14}$$

Because Klaassen observes a closed regional system, it is true that the total production of industry k equals the total demand for the products of this industry. In symbols

$$q^k_* = d^k_{**} \qquad (k = 1, \ldots, M) \tag{6.15}$$

From (6.3) and (6.14) it follows that

$$q_*^k = \frac{\alpha_{**}^{lk}}{\beta_{**}^{lk}} q_*^l \qquad (k, l = 1, \ldots, M) \qquad (6.16)$$

From (6.12), (6.15) and (6.16), (6.13) results.

In selecting industries that could contribute to the development of retarded areas within a regional system, it is important to study the existing industrial structure, i.e. the existing interindustry relations and the regional distribution of total productions. That is why the attraction equation (6.13) describes the relation between fractions produced and demanded in region i.

The coefficients λ (k) and λ (l,k) are denoted as the *demand-attraction coefficient* and the *supply-attraction coefficient* for input factor l. From (6.10) and (6.11) it can be concluded that

$$0 < \lambda(k) < 1 \ ; \ 0 < \lambda(l,k) < 1 \ ; \ \lambda(k) + \sum_{l \neq t} \lambda(l,k) = 1 \qquad (6.17)$$

Owing to Klaassen's assumption with respect to τ_{iw}^k and τ_{wi}^{lk}, none of these coefficients is region specific.

Location theory, attraction theory, and attraction models

In classical location theory, the objective is to determine the optimum location of an *establishment*, that is to say 'a site'. For industries – often comprising several establishments – the place of location is not 'a site' but a *zone*. Klaassen calls his theory to explain the distribution of the total production of a given industry among the zones constituting the regional system, a *location theory*. Van Wickeren, in his doctoral thesis 'Interindustry relations: some attraction models' (1971), gives to the location theory developed by Klaassen the name of *attraction theory* [9] '. . . the theory tries to explain the location of all activities, [10] including the communication sectors'. Moreover, he mentions the equations [11]

$$c_i^k \triangleq \tau_{iw}^k (q_i^k - d_{*_i}^k) + \sum_{l \neq t} \tau_{wi}^{lk} (\beta_{**}^{lk} q_i^k - \alpha_{**}^{lk} q_i^l) \qquad (6.1)'$$

and

$$c_i^k \triangleq \alpha_{**}^{tk} q_i^t \qquad (6.7)$$

'the specific [12] system of structural equations dealing with communication costs' [12] or 'the specific attraction model'. In his line of thought, there are M such models per region.

On the one hand the M attraction equations (6.12), all of them derived from one specific attraction model, are denoted by him as 'reduced form equations of the specific attraction models', on the other hand he refers to them as 'the separate equations of the *general attraction model*'; [14] the latter is valid for region i. Although Van Wickeren indicates for neither type of attraction model which of their variables are to be looked upon as endogenous and which as exogenous, it can be inferred from the equation in matrix notation he presents as 'basic solution of the general attraction model', [15] which of the variables are exogenous. To obtain that basic solution, $d^k_{*_i}$ is written as

$$d^k_{*_i} \equiv \sum_l r^{kl}_{*_i} + f^k_{*_i} \qquad (k = 1, \ldots, M) \qquad (6.18)$$

The symbols $r^{kl}_{*_i}$ and $f^k_{*_i}$ have the following meaning:

$r^{kl}_{*_i}$ the intermediate demand of industry l in region i for products of industry k, located within the regional system;

$f^k_{*_i}$ the final demand [16] in region i for products of industry k, located within the regional system.

By definition it is true that

$$r^{kl}_{*_i} \triangleq rr^{kl}_{*_i} \qquad (k,l = 1, \ldots, M) \qquad (6.19)$$

So the intermediate demand of industry l (in region i) for products of industry k is by definition equal to the supply by industry k to industry l in region i.

Successive substitution of (6.4) into (6.19), (6.19) into (6.18), and (6.18) into (6.12) leads to

$$q^k_i = \sum_l \left(\lambda\,(k)\, \beta^{kl}_{**} + \lambda\,(l,k) \frac{a^{lk}_{**}}{\beta^{lk}_{**}} \right) q^l_i + \lambda\,(k)\, f^k_{*_i} \quad (k = 1, \ldots, M) \qquad (6.20)$$

Equation (6.20) can be written with the help of matrices and vectors, as:

$$q_i = \Omega q_i + \Lambda f_i \qquad (6.21)$$

where

$$(q_i)' \equiv (q_i^1, \ldots, q_i^M) \text{ en } (f_i)' \equiv (f_{*_i}^1, \ldots, f_{*_i}^M) \qquad (6.22)$$

and Ω is a square matrix of order M x M in which the characteristic element ω_{lk} is equal to

$$\omega_{lk} = \lambda(k)\,\beta_{**}^{kl} + \lambda(l,k)\,\frac{\alpha_{**}^{lk}}{\beta_{**}^{lk}} \qquad (6.23)$$

and Λ representing a diagonal matrix of order M x M with on the main diagonal the elements $\lambda(k)$, k = 1, . . . , M. From (6.21) it follows:

$$q_i = (I - \Omega)^{-1} \Lambda f_i \qquad (6.24)$$

Equation (6.24) is the basic solution [17] of the general attraction model (6.12). On that ground we can state that $f_{*_i}^k$ (k = 1, . . . , M) are exogenous variables, and conclude that in (6.12), q_i^k and $d_{*_i}^k$ (k = 1, . . . , M) are to be considered endogenous variables. That conclusion gives rise to the question whether the general attraction model is a structural model, [18] i.e. a 'complete system [19] of equations which describe the structure of the relationship of the economic variables'. The general attraction model consists of M equations and counts 2M endogenous variables; so it is not complete and therefore not a structural model. And that means that the reduced form [20] of the general attraction model is not defined. The same holds for the reduced form of each of the M specific attraction models consisting of equations (6.1)' and (6.7). The specific attraction model for industry k in region i is composed of two equations and M + 2 endogenous variables (q_i^l for l = 1, . . . , M, plus $d_{*_i}^k$ and c_i^k). The M attraction equations (6.12) may not, therefore, be indicated as 'the reduced form equations of the specific attraction models'. Via a different approach, Deacon comes to the same conclusion: '. . . there are endogenous variables on the right-hand side of the equation, so that reduced form is an inappropriate description'. [21]

Although Klaassen, in his study quoted in the introduction, does not use the concept of attraction models, the obvious thing to do now is to check whether the equations he developed, viz. (6.1), (6.2), (6.4), (6.5), and (6.7), as the mathematical formulation of the attraction theory, form a complete system. The following variables are endogenous:

	c^k_i	q^k_i	$d^k_{*_i}$	rr^{lk}_{wi}	$rr^{lk}_{*_i}$	rr^{lk}_{ii}
Number	M	M	M	M^2	M^2	M^2

The total number of endogenous variables evidently amounts to $3M + 3M^2$. The number of equations being $2M + 3M^2$, the system is not complete. By adding equations (6.18) and (6.19) to it, we obtain a larger system that is, indeed, complete. As a new endogenous variable, $r^{kl}_{*_i}$ is introduced, bringing the total number of endogenous variables up to $3M + 4M^2$. The total number of equations also amounting to $3M + 4M^2$ after the addition of (6.18) and (6.19), completeness is a fact. The larger system is called a *structural attraction model.* In the structural attraction model consisting of the equations (6.1), (6.2), (6.4), (6.5), (6.7), (6.18), and (6.19), the variables $f^k_{*_i}$ for k = 1, . . . , M, are exogenous, while τ^k_{iw}, τ^{lk}_{wi}, β^{lk}_{**}, a^{lk}_{**}, and a^{tk}_{**} represent structural parameters (constants). This structural attraction model is valid for region i; it is, therefore, an *intraregional* structural model. It is not an intra-industry (= specific) model, but an *inter-industry* one, as appears from equation (6.18).

Interregional attraction models for a closed regional system

Introduction

If industry k in region i sales its whole production within a regional system consisting of N regions, that production may, for example, be defined as:

$$q^k_i \triangleq \sum_j \left(\sum_l rr^{kl}_{ij} + ff^k_{ij}\right) \qquad (k = 1, \ldots, M; i = 1, \ldots, N) \qquad (6.25)$$

The symbols used on the right-hand side of (6.25) have the following meaning:

rr^{kl}_{ij} intermediate supply by industry k in region i to industry l in region j;

ff^k_{ij} final supply by industry k in region i to region j.

Equation (6.25) can also be written as:

$$q_i^k \triangleq \sum_l rr_{i*}^{kl} + ff_{i*}^k \qquad (k = 1, \ldots, M; i = 1, \ldots, N) \qquad (6.26)$$

where

$$rr_{i*}^{kl} \triangleq \sum_j rr_{ij}^{kl} \quad \text{en} \quad ff_{i*}^k \triangleq \sum_j ff_{ij}^k$$

The former structural attraction model has (6.25) for its basis, the latter (6.26).

Model I

We want to introduce two hypotheses, viz.

$$rr_{ij}^{kl} = \phi(j)\, r_{*j}^{kl} \exp(-\gamma_i^k c_{ij}^{kl}) \quad (k,l = 1, \ldots, M; i,j = 1, \ldots, N) \qquad (6.27)$$

and

$$ff_{ij}^k = \tilde{\phi}(j)\, f_{*j}^k \exp(-\tilde{\gamma}_i^k \tilde{c}_{ij}^k) \quad (k = 1, \ldots, M; i,j = 1, \ldots, N) \qquad (6.28)$$

Equation (6.27) is a causal relationship indicating that the quantities supplied by industry k in region i to industry l in region j depends on the intermediate k-demand of industry l in region j as discounted across space by industry k. In this equation, $\phi(j)$ is a coefficient indicating the accessibility [22] of region j.

Industry k in region i discounts the effective k-demand of industry l in region j, r_{*j}^{kl}, by dividing this demand by the rate of discount $\exp(\gamma_i^k c_{ij}^{kl})$. The symbol c_{ij}^{kl} stands for:

c_{ij}^{kl} the communication costs involved in supplying a unit of the production of industry k in region i to industry l in region j, imposing on industry k in region i. [23]

Coefficient γ_i^k represents the sensitivity of industry k in region i to c_{ij}^{kl}. We assume that this coefficient is independent of both j and l.

Equation (6.28) explains the final supply from industry k in region i to region j. This equation has the same structure as (6.27). The symbol $\tilde{c}_{ij}^k$ stands for the communication involved in the final supply of a unit of product of industry k in region i to region j.

Substitution of (6.27) and (6.28) into (6.25) leads to:

$$q_i^k = \sum_l \sum_j \phi(j)\, r_{*_j}^{kl} \exp(-\gamma_i^k c_{ij}^{kl}) + \sum_j \tilde{\phi}(j)\, f_{*_j}^k \exp(-\tilde{\gamma}_i^k \tilde{c}_{ij}^k)$$

$$(k = 1, \ldots, M;\ i = 1, \ldots, N) \qquad (6.29)$$

The expression

$$\sum_j \phi(j)\, r_{*_j}^{kl} \exp(-\gamma_i^k c_{ij}^{kl})$$

represents the *intermediate l-demand potential* of industry k in region i; the final-demand potential of industry k in region i is represented by

$$\sum_j \tilde{\phi}(j)\, f_{*_j}^k \exp(-\tilde{\gamma}_i^k \tilde{c}_{ij}^k)$$

The production of industry k in region i is evidently explained by the M intermediate-demand potentials and the final-demand potential of this industry. [24]

We assume that the intermediate k-demand of industry l in region j is provided for exclusively by supplies from industry k located within the regional system. A similar assumption is made with respect to $f_{*_j}^k$, so that

$$r_{*_j}^{kl} \triangleq rr_{*_j}^{kl} \text{ and } f_{*_j}^k \triangleq ff_{*_j}^k \qquad (k,l = 1, \ldots, M;\ j = 1, \ldots, N) \qquad (6.30)$$

Summation of (6.27) and (6.28) over i gives

$$rr_{*_j}^{kl} = \phi(j)\, r_{*_j}^{kl} \sum_i \exp(-\gamma_i^k c_{ij}^{kl}) \quad (k,l = 1, \ldots, M;\ j = 1, \ldots, N) \qquad (6.31)$$

and

$$ff_{*_j}^k = \tilde{\phi}(j)\, f_{*_j}^k \sum_i \exp(-\tilde{\gamma}_i^k \tilde{c}_{ij}^k) \quad (k = 1, \ldots, M;\ j = 1, \ldots, N) \qquad (6.32)$$

From (6.30) through (6.32) it follows that

$$\phi(j) = \frac{1}{\sum_i \exp(-\gamma_i^k c_{ij}^{kl})} \text{ and } \tilde{\phi}(j) = \frac{1}{\sum_i \exp(-\tilde{\gamma}_i^k \tilde{c}_{ij}^k)} \qquad (6.33)$$

Analogously with (6.4) we write $rr^{kl}_{*_j}$ as

$$rr^{kl}_{*_j} \triangleq \beta^{kl}_{**} q^l_j \qquad (k,l = 1, \ldots, M; j = 1, \ldots, N) \qquad (6.34)$$

Combination of (6.29), (6.30), (6.33), and (6.34) produces the *structural equations* of model I:

$$q^k_i = \sum_l \beta^{kl}_{**} \sum_j q^l_j \frac{\exp(-\gamma^k_i c^{kl}_{ij})}{\sum_i \exp(-\gamma^k_i c^{kl}_{ij})} + \sum_j f^k_{*_j} \frac{\exp(-\tilde{\gamma}^k \tilde{c}^k_{ij})}{\sum_i \exp(-\tilde{\gamma}^k_i \tilde{c}^k_{ij})}$$

$$(k = 1, \ldots, M; i = 1, \ldots, N) \qquad (6.35)$$

In this system, variables q^k_i $(k = 1, \ldots, M; i = 1, \ldots, N)$ are endogenous. Exogenous are $f^k_{*_j}$, c^{kl}_{ij}, and $\tilde{c}^k_{ij}$ $(k,l = 1, \ldots, M;$ $i,j = 1, \ldots, N)$. The system is complete, the number of equations being equal to the number of endogenous variables.

From (6.35) it is evident that the production of any industry in any region is influenced both by the productions of all industries in all regions and by the final demand in all regions for the products of the industry in question. Moreover, the production of that industry depends on the communication costs it incurs for

- the supply of one unit of output to industry 1, . . . , M in region 1, . . . , N;
- the final supply of one unit of output to region 1, . . . , N.

Model I is a *system of simultaneous equations* [25] and can be denoted as a *structural interregional attraction model.*

To conclude this section three marginal notes:

1. The practical significance of the non-linear system (6.35) is small as long as no data [26] are available for c^{kl}_{ij} and $\tilde{c}^k_{ij}$, for without such data we cannot estimate the parameters of model I.
2. Model I is a deterministic model, that is to say the model has been built up out of exact functional relations. However, 'exact structural relations are admittedly unrealistic'. [27] That difficulty can be remedied by adding to relations (6.27) and (6.28) a *random variable* ϵ representing the combined effect of all neglected factors.

3. A theoretical drawback attaching to model I will be discussed after model II has been introduced; see 'Practical and theoretical objections to models I and II', p.127.

Model II

(6.27) and (6.28) are the relations underlying model I. With the help of these relations the intermediary supply from industry k in region i to industry l in region j and the final supply from the former industry to region j is explained. The 'direction' of the supply is, therefore, given a priori. With model II it is not so, however; the first concern with this model is to explain the intermediary supply from industry k in region i to industry l in the regional system (rr^{kl}_{i*}) and the final supply from the former industry to all regions of the regional system together (ff^{k}_{i*}).

We hypothesise that

- rr^{kl}_{i*} depends on the production of industry l in all regions of the regional system and the communication costs involved in the supply of one unit produced by industry k in region i to industry l in region 1, . . . , N;
- ff^{k}_{i*} is determined by the final demand in region 1, . . . , N for the products of industry k, established within the regional system and the communication costs involved in the final supply of one unit of industry k in region i to region 1, . . . , N.

In symbols: [28]

$$rr^{kl}_{i*} = \Phi_k (q^l, c^{kl}_i) + \epsilon^{kl}_{i*} \quad (k,l = 1, \ldots, M; i = 1, \ldots, N) \quad (6.36)$$

and

$$ff^{k}_{i*} = \tilde{\Phi}_k (f^k_*, \tilde{c}^k_i) + \tilde{\epsilon}^{k}_{i*} \quad (k = 1, \ldots, M; i = 1, \ldots, N) \quad (6.37)$$

with

$$(q^l)' \triangleq (q^l_1, \ldots, q^l_j, \ldots, q^l_N) \quad (6.38)$$

$$(f^k_*)' \triangleq (f^k_{*1}, \ldots, f^k_{*j}, \ldots, f^k_{*N}) \quad (6.39)$$

$$(c^{kl}_i)' \triangleq (c^{kl}_{i1}, \ldots, c^{kl}_{ij}, \ldots, c^{kl}_{iN}) \quad (6.40)$$

$$(\tilde{c}^k_i)' \triangleq (\tilde{c}^k_{i1}, \ldots, \tilde{c}^k_{ij}, \ldots, \tilde{c}^k_{iN}) \quad (6.41)$$

Here the symbols Φ_k and $\tilde{\Phi}_k$ denote unknown scalar functions of the variables shown in parentheses. In (6.36) and (6.37), ϵ^{kl}_{i*} and $\tilde{\epsilon}^{kl}_{i*}$ represent disturbances [29] to be considered as random variables.

In view of the lack of data about c^{kl}_i and $\tilde{c}^k_i$ we observe

$$\Phi_k (q^l, \hat{c}^{kl}_i) = \Phi_k (q^l, c^{kl}_i) \Big]_{c^{kl}_i = \hat{c}^{kl}_i} \tag{6.42}$$

in which $\hat{c}^{kl}_i$ represents a vector of values adopted by c^{kl}_i and

$$\tilde{\Phi}_k (f^k_*, \hat{\tilde{c}}^k_i) = \tilde{\Phi}_k (f^k_*, \tilde{c}^k_i) \Big]_{\tilde{c}^k_i = \hat{\tilde{c}}^k_i} \tag{6.43}$$

where $\hat{\tilde{c}}^k_i$ represents a vector of values adopted by $\tilde{c}^k_i$.

We replace the left-hand side of (6.42) by the expansion of Φ_k in a Taylor series around the point

$$(E (q^l_1), \ldots, E (q^l_j), \ldots, E (q^l_N))$$

where the q^l_j assume their expected value $E (q^l_j)$, $j = 1, \ldots, N$.

Applying Taylor's theorem for $n = 1$ we get [30]

$$\Phi_k (q^l, \hat{c}^{kl}_i) = \Phi_k (E (q^l), \hat{c}^{kl}_i) + \sum_j (q^l_j - E (q^l_j)) \Psi_{kj} (E (q^l), \hat{c}^{kl}_i) + R^{(1)}_{kli} \tag{6.44}$$

where

$$\frac{\partial \Phi_k}{\partial_q{}^l} \triangleq \begin{bmatrix} \dfrac{\partial \Phi_k}{\partial q^l_1} \\ \vdots \\ \dfrac{\partial \Phi_k}{\partial q^l_j} \\ \vdots \\ \dfrac{\partial \Phi_k}{\partial q^l_N} \end{bmatrix} = \begin{bmatrix} \Psi_{kl} (q^l, \hat{c}^{kl}_i) \\ \vdots \\ \Psi_{kj} (q^l, \hat{c}^{kl}_i) \\ \vdots \\ \Psi_{kN} (q^l, \hat{c}^{kl}_i) \end{bmatrix} \tag{6.45}$$

$\Phi_k\,(E\,(q^l), \hat{c}_i^{kl})$ is for a given k and l a variable coefficient, in the sense that the value is determined by $\hat{c}_i^{kl}$. We write

$$\phi_{kl}\;(\hat{c}_i^{kl}) = \Phi_k\,(E\,(q^l), \hat{c}_i^{kl}) \tag{6.46}$$

The partial derivative $\Psi_{kj}\,(E\,(q^l), \hat{c}_i^{kl})$ is for given k, l and j also a variable coefficient, the value of which is determined by $\hat{c}_i^{kl}$. So, we may write

$$\psi_{klj}\;(\hat{c}_i^{kl}) = \Psi_{kj}\,(E\,(q^l), \hat{c}_i^{kl}) \tag{6.47}$$

Next we introduce

$$\phi_{*kl}\;(\hat{c}_i^{kl}) \triangleq \phi_{kl}\,(\hat{c}_i^{kl}) - \sum_j E\,(q_j^l)\;\psi_{klj}\;(\hat{c}_i^{kl}) \tag{6.48}$$

Applying (6.46) through (6.48) we can write (6.44) as

$$\Phi_k\;(q^l, \hat{c}_i^{kl}) = \phi_{*kl}\;(\hat{c}_i^{kl}) + \sum_j q_j^l\;\psi_{klj}\;(\hat{c}_i^{kl}) + R_{kli}^{(1)} \tag{6.49}$$

in which $R_{kli}^{(1)}$ represents the difference, or 'remainder', between Φ_k and the series written.

Similarly as with (6.44) it can be derived that

$$\tilde{\Phi}_k\;(f_*^k, \hat{\tilde{c}}_i^k) = \tilde{\Phi}_k\,(\bar{f}_*^k, \hat{\tilde{c}}_i^k) + \sum_j (f_{*j}^k - \bar{f}_{*j}^k)\,\tilde{\Psi}_{kj}\,(\bar{f}_*^k, \hat{\tilde{c}}_i^k) + \tilde{R}_{ki}^{(1)} \tag{6.50}$$

with the expansion of $\tilde{\Phi}_k$ in a Taylor series occurring around the fixed point

$(\bar{f}_{*1}^k, \ldots, \bar{f}_{*j}^k, \ldots, \bar{f}_{*N}^k)$

where the f_{*j}^k assume [31] the value $\bar{f}_{*j}^k$, $j = 1, \ldots, N$.

Analogously to (6.45) it is true that

$$\frac{\partial \tilde{\Phi}_k}{\partial f_*^k} \triangleq \begin{bmatrix} \frac{\partial \tilde{\Phi}_k}{\partial f_{*1}^k} \\ \vdots \\ \frac{\partial \tilde{\Phi}_k}{\partial f_{*j}^k} \\ \vdots \\ \frac{\partial \tilde{\Phi}_k}{\partial f_{*N}^k} \end{bmatrix} = \begin{bmatrix} \tilde{\Psi}_{k1} \ (f_*^k, \hat{\tilde{c}}_i^k) \\ \vdots \\ \tilde{\Psi}_{kj} \ (f_*^k, \hat{\tilde{c}}_i^k) \\ \vdots \\ \tilde{\Psi}_{kN} \ (f_*^k, \hat{\tilde{c}}_i^k) \end{bmatrix} \tag{6.51}$$

By introducing

$$\tilde{\phi}_k \ (\hat{\tilde{c}}_i^k) = \tilde{\Phi}_k \ (\bar{f}_*^k, \hat{\tilde{c}}_i^k) \tag{6.52}$$

$$\psi_{kj} \ (\hat{\tilde{c}}_i^k) = \tilde{\Psi}_{kj} \ (\bar{f}_*^k, \hat{\tilde{c}}_i^k) \tag{6.53}$$

and

$$\tilde{\phi}_k^* \ (\hat{\tilde{c}}_i^k) = \tilde{\phi}_k \ (\hat{\tilde{c}}_i^k) - \sum_j \bar{f}_{*j}^k \ \tilde{\psi}_{kj} \ (\hat{\tilde{c}}_i^k) \tag{6.54}$$

(6.50) can be written as

$$\tilde{\Phi}_k \ (f_*^k, \hat{\tilde{c}}_i^k) = \tilde{\phi}_k^* \ (\hat{\tilde{c}}_i^k) + \sum_j f_{*j}^k \ \tilde{\psi}_{kj} \ (\hat{\tilde{c}}_i^k) + \tilde{R}_{ki}^{(1)} \tag{6.55}$$

'The purpose of estimation (of the parameters of a model) is almost always prediction, directly or ultimately, of the values of endogenous variables'. [32] The same holds in our case; we want to predict the volume of the productions of all industries in all regions of a regional system. But apart from that we want to understand the structure of the regional system so that we can supply materials for a regional industrialisation policy; such a policy cannot be pursued effectively unless there is sufficient insight into the interwovenness of the industries. That means that we want to have at our disposal not only the estimates of the reduced form parameters — with a view to forecasting the endogenous variables — but also those of the structural parameters — in view of the interwovenness of the industries.

To estimate all structural parameters of a model, it should be identified, [33] and observations on the current endogenous and predetermined [34] variables should be available. Assuming, for simplicity's sake, that the endogenous and exogenous variables are not lagged, we operate the vector symbols

$$y' \triangleq (y'_1, \ldots, y'_t, \ldots, y'_T);\ y'_t \triangleq (y_{1t}, \ldots, y_{gt}, \ldots, y_{Gt})$$

and

$$x' \triangleq (x'_1, \ldots, x'_t, \ldots, x'_T);\ x'_t \triangleq (x_{1t}, \ldots, x_{ht}, \ldots, x_{Ht})$$

for the set of respectively the observations on endogenous and exogenous variables. The subscript t refers to the time point at which, or period for which, the variable in question is measured.

In our case we have at our disposal, for estimating the structural parameters of model II – still to be formulated – the vectors **y** and **x**, which consist only of observations for one given t, say $\hat{t}$. The situation is not extreme: 'As compared to traditional (intertemporal) econometrics, spatial econometrics is beset with the difficulty that more often than not time series are not available to estimate the relevant parameters of a model'. [35] In concrete terms vectors **y** and **x** read, in relation to model II:

$$y' = y'_{\hat{t}} = \hat{q}' \triangleq ((\hat{q}^1)', \ldots, (\hat{q}^k)', \ldots, (\hat{q}^M)') \qquad (6.56)$$

where

$$(\hat{q}^k)' \triangleq (\hat{q}^k_1, \ldots, \hat{q}^k_i, \ldots, \hat{q}^k_N) \qquad (6.57)$$

and

$$x' = x'_{\hat{t}} = \hat{f}' \triangleq ((\hat{f}^1)', \ldots, (\hat{f}^k)', \ldots, (\hat{f}^M)') \qquad (6.58)$$

where

$$(\hat{f}^k)' \triangleq (\hat{f}^k_{*1}, \ldots, \hat{f}^k_{*i}, \ldots, \hat{f}^k_{*N}) \qquad (6.59)$$

Defining $\hat{c}^{kl}_{ij}$ and $\hat{\tilde{c}}^k_{ij}$ more closely as an average value assumed by c^{kl}_{ij} and $\tilde{c}^k_{ij}$, respectively, over period $\hat{t}$, we find from (6.46) through (6.48) and (6.52) through (6.54) that the structural parameters in (6.49) and (6.55) are not independent of t. Hence, model II is to be observed only in relation to period $\hat{t}$. Because

$$\hat{q}^k_i \triangleq \sum_l \hat{rr}^{kl}_{i*} + \hat{ff}^k_{i*} \quad (k = 1, \ldots, M;\ i = 1, \ldots, N) \qquad (6.26)'$$

$$\hat{rr}^{kl}_{i*} = \Phi_k(\hat{q}^l, \hat{c}^{kl}_i) + \hat{\epsilon}^{kl}_{i*} \quad (k, l = 1, \ldots, M; i = 1, \ldots, N) \quad (6.36)'$$

$$\hat{ff}^{k}_{i*} = \tilde{\Phi}_k(\hat{f}^k_*, \hat{\tilde{c}}^k_i) + \hat{\tilde{\epsilon}}^k_{i*} \quad (k = 1, \ldots, M; i = 1, \ldots, N) \quad (6.37)'$$

$$\Phi_k(\hat{q}^l, \hat{c}^{kl}_i) = \phi^*_{kl}(\hat{c}^{kl}_i) + \sum_j \hat{q}^l_j \, \psi_{klj}(\hat{c}^{kl}_i) + \hat{R}^{(1)}_{kli} \quad (6.49)'$$

and

$$\tilde{\Phi}_k(\hat{f}^k_*, \hat{\tilde{c}}^k_i) = \tilde{\phi}^*_k(\hat{\tilde{c}}^k_i) + \sum_j \hat{f}^k_{*j} \, \tilde{\psi}_{kj}(\hat{\tilde{c}}^k_i) + \hat{\tilde{R}}^{(1)}_{ki} \quad (6.55)'$$

we will formulate model II for period $\hat{t}$ as follows: [36]

$$\boxed{\hat{q}^k_i = \phi^{**}_{ki} + \sum_l \sum_j \psi_{kilj} \, \hat{q}^l_j + \sum_j \tilde{\psi}_{kij} \, \hat{f}^k_{*j} + \hat{\mu}^k_i \quad (k = 1, \ldots, M; i = 1, \ldots, N)} \quad (6.60)$$

where

$$\phi^{**}_{ki} = \sum_l \phi^*_{kl}(\hat{c}^{kl}_i) + \tilde{\phi}^*_k(\hat{\tilde{c}}^k_i)$$

$$\psi_{kilj} = \psi_{klj}(\hat{c}^{kl}_i) \text{ and } \tilde{\psi}_{kij} = \tilde{\psi}_{kj}(\hat{\tilde{c}}^k_i)$$

Moreover it is true that

$$\hat{\mu}^k_i = \sum_l \hat{\epsilon}^{kl}_{i*} + \hat{\tilde{\epsilon}}^k_{i*} + \sum_l \hat{R}^{(1)}_{kli} + \hat{\tilde{R}}^{(1)}_{ki}$$

In the interregional attraction model II, variables $\hat{q}^k_i$ (k = 1, . . . , M; i = 1, . . . , N) are endogenous, and $\hat{f}^k_{*j}$ (k = 1, . . . , M; j = 1, . . . , N) exogenous. The number of equations and the number of endogenous variables both equal MN. Model II for period $\hat{t}$ is complete, then. In this model too – compare (6.35) – the production of an industry in a region depends on both the productions of all industries in all regions and the final demand in all regions for the products of the industry concerned. However, c^{kl}_{ij} and $\tilde{c}^k_{ij}$ (l = 1, . . . , M; i,j = 1, . . . , N) are missing from model II as explanatory variables for q^k_i.

With the help of coefficient matrices

$$A \triangleq \begin{bmatrix} A_{11} & A_{12} & \cdots & A_{1M} \\ A_{21} & A_{22} & \cdots & A_{2M} \\ \vdots & \vdots & \ddots & \vdots \\ A_{M1} & A_{M2} & \cdots & A_{MM} \end{bmatrix}, \quad A_{kk} \triangleq \begin{bmatrix} 1-\psi_{k1k1} & -\psi_{k1k2} & \cdots & -\psi_{k1kN} \\ -\psi_{k2k1} & 1-\psi_{k2k2} & \cdots & -\psi_{k2kN} \\ \vdots & \vdots & \ddots & \vdots \\ -\psi_{kNk1} & -\psi_{kNk2} & \cdots & 1-\psi_{kNkN} \end{bmatrix}$$

$$\underset{(k \neq l)}{A_{kl}} \triangleq \begin{bmatrix} -\psi_{k1l1} & -\psi_{k1l2} & \cdots & -\psi_{k1lN} \\ -\psi_{k2l1} & -\psi_{k2l2} & \cdots & -\psi_{k2lN} \\ \vdots & \vdots & \ddots & \vdots \\ -\psi_{kNl1} & -\psi_{kNl2} & \cdots & -\psi_{kNlN} \end{bmatrix}, \quad B \triangleq \left[\begin{array}{c|cccc} b_{01} & B_1 & 0 & \cdots & 0 \\ b_{02} & 0 & B_2 & \cdots & 0 \\ \vdots & \vdots & \vdots & \ddots & \vdots \\ b_{0M} & 0 & 0 & \cdots & B_M \end{array}\right]$$

$$B_k \triangleq \begin{bmatrix} -\tilde{\psi}_{k11} & -\tilde{\psi}_{k12} & \cdots & -\tilde{\psi}_{k1N} \\ -\tilde{\psi}_{k21} & -\tilde{\psi}_{k22} & \cdots & -\tilde{\psi}_{k2N} \\ \vdots & \vdots & \ddots & \vdots \\ -\tilde{\psi}_{kN1} & -\tilde{\psi}_{kN2} & \cdots & -\tilde{\psi}_{kNN} \end{bmatrix}, \quad b_{ok} \triangleq \begin{bmatrix} -\phi^{**}_{k1} \\ -\phi^{**}_{k2} \\ \vdots \\ -\phi^{**}_{kN} \end{bmatrix}, \quad \hat{f}^* = \begin{bmatrix} \frac{1}{\hat{f}} \end{bmatrix}$$

the structural equations (6.60) can be written in matrix form as

$$A\,\hat{q} + B\,\hat{f}^* = \hat{\mu} \tag{6.61}$$

vector $\hat{\mu}$ having the same structure as $\hat{q}$.

Practical and theoretical objections to models I and II

At the end of section 'Model I', p.118, it was remarked that model I has so far been of little importance since no data are as yet available for the variables c^{kl}_{ij} and $\tilde{c}^{k}_{ij}$, so that the parameters of the model cannot yet be estimated. Now that we have succeeded in formulating a model for which the data problems as to c_{ij}^{kl} and $\tilde{c}_{ij}^{k}$ do not present themselves, one is inclined to conclude that the parameters of model II can, indeed, be estimated.

Such a conclusion would not be right, however. To explain q_i^k without taking c_{ij}^{kl} and $\tilde{c}_{ij}^{k}$ into consideration as explanatory variables, we have to pay the price of a loss in identification state: while model I is overidentified, model II is underidentified. As a result it is impossible to estimate all the structural parameters of model II and thus to determine the structure of the regional system, a fact that, to our minds, greatly reduces the model's practical significance.

Model I is identified because each equation is identified, owing to the nonlinear character of all equations in the system (6.35). [37] It is at the same time overidentified because all its equations are overidentified; indeed, of each separate equation it is true that 'the total number of variables in the model' (= $2MN + MN^2 + M^2N^2$) minus 'the number of variables included in a particular equation' (= $MN + MN^2 + N + N^2$) is greater than 'the total number of equations minus one' (= $MN - 1$). [38]

In general a model can be said to be underidentified when one or more of its equations are underidentified. The individual equations of model II do not satisfy the order condition for identification because 'the total number of variables in the model' (= $2\,MN + 1$) minus 'the number of variables included in a particular equation' (= $MN + N + 1$) is smaller than 'the total number of equations minus one' (= $MN - 1$). [39] Therefore, model II is underidentified.

So, even if we assume that a statistician has data on all of the variables q_i^k and f_{*j}^k, we find that the structural parameters of

- model I cannot be estimated because data on the variables c_{ij}^{kl} and $\tilde{c}_{ij}^{k}$ are lacking; and
- those of model II cannot be estimated because this model is underidentified.

Now that it has been verified that model I is identified, the problem that its parameters cannot (yet) be estimated will be left out of account in the second part of the present section; instead we shall give some more attention to the fact that this model has been designed for a closed regional system. In a closed regional system consisting of N regions:

- the production of industry k in region i is sold within the regional system;

– the intermediate k-demand of industry l in region i is provided for by supplies from industry k in region 1, . . . , N;
– the final k-demand in region j is met by supplies from industry k in region 1, . . . , N.

On the assumption that industry k in region i makes its deliveries exclusively from its current production – and not from stocks built up in previous periods – the total production of industry k, located within the regional system, is equal to the sum of intermediate and final k-demand in all regions of the system together. A theoretical drawback of model I is that it cannot be proved for any of the industries that the quality is exactly achieved. From (6.35) it follows in fact that

$$\sum_i q_i^k = \sum_l \sum_j \beta_{**}^{kl} \; q_j^l + \sum_j f_{*j}^k \qquad (k = 1, \ldots, M) \qquad (6.62)$$

In words: the total production of industry k, located within the regional system, is equal to the sum of the intermediate supplies of that industry and the final k-demand within the system.

So, the above mentioned equality holds under the condition that the intermediate supplies equal intermediate k-demand within the system, that is to say

$$\sum_l \sum_j rr_{*j}^{kl} = \sum_l \sum_j \beta_{**}^{kl} \; q_j^l = \sum_l \sum_j r_{*j}^{kl} \qquad (6.63)$$

So, the disadvantage of model I is that it cannot be derived *from this model* that (6.63) is satisfied; consequently, it does not deserve the title of 'interregional attraction model for a *closed* regional system'.

Models III and IV

The properties of a closed regional system and the assumption that industry k in region i makes its deliveries exclusively from the current production, can also be expressed by means of symbols. Of a closed regional system consisting of N regions, it can be stated that:

1. $$q_i^k \triangleq \sum_j (\sum_l rr_{ij}^{kl} + ff_{ij}^k) \qquad (k = 1, \ldots, M; i = 1, \ldots, N) \qquad (6.25)$$

2. $$\sum_i rr_{ij}^{kl} = r_{*j}^{kl} \qquad (k,l = 1, \ldots, M; j = 1, \ldots, N) \qquad (6.64)$$

3. $$\sum_i ff_{ij}^k = f_{*j}^k \qquad (k = 1, \ldots, M; j = 1, \ldots, N) \qquad (6.65)$$

By including (6.25), (6.64) and (6.65) as *structural equations* in an interregional attraction model for a closed regional system, the theoretical objection to model I is remedied. For it follows from these equations that

$$\sum_i q_i^k = \sum_j \sum_l (\sum_i rr_{ij}^{kl}) + \sum_j \sum_i ff_{ij}^k = \sum_j \sum_l r_{*j}^{kl} + \sum_j f_{*j}^k \quad (k = 1, \ldots, M)$$

In words: the total production of industry k, located within the regional system, is equal to the sum of intermediate and final k-demand in all regions of the system together.

Now model III is composed of (6.25), (6.64) and (6.65), and the causal relationships (6.66) through (6.68) to be formulated below.

$$rr_{ij}^{kl} = \frac{r_{*j}^{kl} \exp(-\gamma_i^k c_{ij}^{kl} + \epsilon_{ij}^{kl})}{\sum_j r_{*j}^{kl} \exp(-\gamma_i^k c_{ij}^{kl} + \epsilon_{ij}^{kl})} \cdot rr_{i*}^{kl}$$

$$(k,l = 1, \ldots M;\ i,j = 1, \ldots, N) \qquad (6.66)$$

$$ff_{ij}^k = \frac{f_{*j}^k \exp(-\tilde{\gamma}_i^k \tilde{c}_{ij}^k + \tilde{\epsilon}_{ij}^k)}{\sum_j f_{*j}^k \exp(-\tilde{\gamma}_i^k \tilde{c}_{ij}^k + \tilde{\epsilon}_{ij}^k)} \cdot ff_{i*}^k$$

$$(k = 1, \ldots, M;\ i,j = 1, \ldots, N) \qquad (6.67)$$

$$r_{*j}^{kl} = \phi_{kl}\, q_j^l + \mu_{*j}^{kl} \qquad (k,l = 1, \ldots, M;\ j = 1, \ldots, N) \qquad (6.68)$$

According to (6.66) the relation between

- the supply from industry k in region i to industry l in region j (rr_{ij}^{kl}),

and

- the supply from the same industry to industry l located within the regional system (rr_{i*}^{kl})

is equal to that between

- the intermediate k-demand of industry l in region j as discounted across space by industry k in region i,

and

– the intermediate l-demand potential of industry k in region i,

provided that disturbance terms, exp (ϵ_{ij}^{kl}), are included in the right-hand side of the equation.

Equation (6.67) describes the distribution of the final supply from industry k in region i (ff_{i*}^{k}) among the regions of the regional system; this equation has the same structure as (6.66).

In (6.68) it is assumed that the intermediate demand of industry l in region j for the products of industry k is a linear function of the production of industry l in region j.

In model III the variables q_i^k, rr_{ij}^{kl}, rr_{i*}^{kl}, r_{*j}^{kl}, ff_{ij}^{k}, and ff_{i*}^{k} are endogenous, and variables f_{*j}^{k}, c_{ij}^{kl} and $\tilde{c}_{ij}^{k}$ exogenous. The number of endogenous variables as well as the number of equations amounts to MN (2 + 2M + N + MN); therefore, the model is complete. Because model III counts linear as well as non-linear relations, every equation has a unique statistical form, that is to say every equation will be identified, [40] and consequently the system of structural equations (6.25), (6.64) through (6.68) is identified too. With the help of the order condition it can be assessed that the equations (6.66) through (6.68) are individually overidentified.

The theoretical objection to model I does not apply to model III, but the practical objection does. The parameters of model III cannot be estimated either for want of data on all of the variables c_{ij}^{kl} and $\tilde{c}_{ij}^{k}$.

To achieve our purpose – forecasting the productions of all industries in all regions of a regional system and determining the structure of the system – we should have available on the one hand an identified model, and on the other observations to estimate this model's parameters. Both are available if in model III

- either the exogenous variables c_{ij}^{kl} and $\tilde{c}_{ij}^{k}$ are replaced with given functions of the *physical* distance between the regions i and j, [41]
- or equation (6.66) is replaced with the combination (6.36), (6.49); and equation (6.67) is substituted by the combination (6.37), (6.55).

In the latter case model IV emerges, composed of the following structural equations: [42]

$$q_i^k \triangleq \sum_l rr_{i*}^{kl} + ff_{i*}^{k} \qquad (k = 1, \ldots, M; i = 1, \ldots, N) \qquad (6.26)$$

$$rr_{*j}^{kl} = r_{*j}^{kl} \qquad (k,l = 1, \ldots, M; j = 1, \ldots, N) \qquad (6.64)'$$

$$ff^{k}_{*_j} = f^{k}_{*_j} \qquad (k = 1, \ldots, M; j = 1, \ldots, N) \qquad (6.65)'$$

$$rr^{kl}_{i*} = \phi^{*}_{kl} \; (\hat{c}^{kl}_{i}) + \sum_{j} q^{l}_{j} \; \psi_{kjl} \; (\hat{c}^{kl}_{i}) + \omega^{kl}_{i*}$$

$$(k,l = 1, \ldots, M; i = 1, \ldots, N) \qquad (6.69)$$

$$ff^{k}_{i*} = \tilde{\phi}^{*}_{k} \; (\hat{\tilde{c}}^{k}_{i}) + \sum_{j} f^{k}_{*_j} \; \tilde{\psi}_{kj} \; (\hat{\tilde{c}}^{k}_{i}) + \tilde{\omega}^{k}_{i*}$$

$$(k = 1, \ldots, M; i = 1, \ldots, N) \qquad (6.70)$$

$$r^{kl}_{*_j} = \phi_{kl} \, q^{l}_{j} + \mu^{kl}_{*_j} \qquad (k,l = 1, \ldots, M; j = 1, \ldots, N) \qquad (6.68)$$

In model IV the variables q^{k}_{i}, rr^{kl}_{i*}, $rr^{kl}_{*_j}$, $r^{kl}_{*_j}$, ff^{k}_{i*} and $ff^{k}_{*_j}$ are endogenous; variable $f^{k}_{*_j}$ is exogenous. The number of equations and the number of endogenous variables are both equal to 3MN (1 + M); hence the model is complete. If it can be shown that model IV is identified, the parameters of this model can, in principle, be estimated. Each of the equations (6.68) through (6.70) individually satisfies the order condition for identification. The question whether or not the rank condition is satisfied too, calls for further investigation.

Conclusions

1. The equations developed by Klaassen as the mathematical formulation of attraction theory do not constitute a complete system of structural equations.
2. The specific and general attraction models formulated by Van Wickeren do not form a structural model.
3. Interregional attraction model I is identified but has a practical and a theoretical drawback. The practical objection refers to the fact that for some variables no data are available, so that the parameters of the model cannot be estimated. The theoretical drawback is that it does not appear from the model that the production of an industry located within a regional system is equal to the sum of the corresponding intermediary and final demands within the system.
4. The practical objection to model I does not apply to model II, but the structural parameters of the latter cannot be estimated either, because the model is not identified.

5 The theoretical objection to model I does not rise with the identified model III, but the practical objection does.

6 The objections to model I do not apply to model IV, but in contrast with model I, the spatial dimension is no longer explicitly present in model IV owing to the lack of discount rates for effective demand. Model IV satisfies the necessary condition for identification; whether it also meets the sufficient condition is a matter for further research.

Notes

[1] Klaassen, L.H., (1967), p.57.
[2] Klaassen, L.H., (1974), p.1. For the contents of the concept of communication costs, see chapter 4, p.54.
[3] Klaassen, L.H., (1967), p.116.
[4] The relevant region for an industry k located in region i may partly lie outside region i.
[5] Klaassen, L.H., (1967), pp 79 and 120.
[6] Klaassen supposes that $\tau^{k}_{iw} = \tau^{k}_{1w}$ en $\tau^{lk}_{wi} = \tau^{lk}_{w1}$ for all i.
[7] Klaassen, L.H., (1967), pp 118 and 120.
[8] Klaassen, L.H., (1967), p.77.
[9] Op.cit., p.9.
[10] 'Industry (in this sense) is almost synonymous with economic activity', op.cit., p.4.
[11] Equation (6.1)′ follows immediately from (6.1) and (6.6).
[12] Specific *to industry k.*
[13] Op.cit., p.4, note 1.
[14] Op.cit., p.4, note 1.
[15] Op.cit., p.13.
[16] Contrary to what is common in input-output analysis, exports are not included in this final demand.
[17] Op.cit., p.13.
[18] Koutsoyiannis (1973), p.326.
[19] 'Mathematical completeness: it requires that the model has as many independent equations as endogenous variables'. Wonnacott and Wonnacott (1970), p.189.
[20] The reduced form of a structural model is defined as 'the model in which the endogenous variables are expressed as a function of the predetermined variables only'. Koutsoyiannis (1973), p.327.
[21] Deacon (1975), p.114.
[22] See section titled 'The attraction equations derived by Klaassen for a closed regional system', p.110.

[23] c_{ij}^{kl} is a function of, among others, the physical distance between regions i and j and the social distance between industries k and l.
[24] See equation (4.5), p.50.
[25] 'A system describing the joint dependence of variables is called a system of simultaneous equations'. Koutsoyiannis (1973), p.321.
[26] See section titled 'The attraction equations derived by Klaassen for a closed regional system', p.110.
[27] Marschak, J., (1953), p.12.
[28] We assume that $q^l \neq 0$ and $f_*^k \neq 0$.
[29] See marginal note on p.120.
[30] The function Φ_k $(q^l, \hat{c}_i^{kl})$ is supposed to have continuous and finite partial derivatives up to any desired order at $(E(q_1^l), \ldots, E(q_j^l), \ldots, E(q_N^l))$.
[31] We assume that f_{*j}^k $(k = 1, \ldots, M; j = 1, \ldots, N)$ is not stochastic.
[32] Koopmans, T.C. and Hood, W.M.C., (1953), p.127.
[33] 'A model is identified if it is in a unique statistical form, enabling unique estimates of its parameters to be subsequently made from sample data', Koutsoyiannis (1973), p.336.
[34] Variables determined either exogenously or in earlier time units (= lagged endogenous).
[35] Van Leeuwen, I., Paelinck, J. and Wagenaar, Sj., (1976), pp 12 and 13.
[36] $(\hat{f}_{*1}^k, \ldots, \hat{f}_{*j}^k, \ldots, \hat{f}_{*N}^k)$ lies in the neighbourhood of $(\bar{f}_{*1}^k, \ldots, \bar{f}_{*j}^k, \ldots, \bar{f}_{*N}^k)$.
[37] Koutsoyiannis, A., (1973), p.354.
[38] The condition $(2MN + MN^2 + M^2N^2) - (MN + MN^2 + N + N^2) > MN - 1$ is satisfied if $N^2 (M^2 - 1) + 1 - N > 0$. Because $M \geqslant 2$ and $N \geqslant 2$, it is true that $M^2 - 1 > 1 \longrightarrow N^2 (M^2 - 1) > N^2 > N \longrightarrow N^2 (M^2 - 1) - N > 0$, and hence also that $N^2 (M^2 - 1) + -N + 1 > 0$.
[39] $N \geqslant 2$.
[40] Koutsoyiannis, A., (1973), p.354.
[41] In carrying out this substitution the fact that c_{ij}^{kl} and $\tilde{c}_{ij}^k$ depend also on *social* distances is neglected.
[42] (6.64)′ and (6.65)′ are equations which have to apply ex ante; equation (6.30), on the contrary, is an ex-post relation.

References

Deacon, David A., *Regional Policy and the Location of Industry: an Application of Attraction Theory,* Durham University, 1975.

Klaassen, Leo H., *Methods of Selecting Industries for Depressed Areas,* OECD, Paris,1967.

Klaassen, Leo H., *Some Further Considerations on Attraction Analysis, Foundations of Empirical Economic Research*, NEI, Rotterdam, 1974.

Koopmans, Tjalling C. and Hood, Wm. C., 'The Estimation of Simultaneous Linear Economic Relationships', *Studies in Econometric Method*, ed. Wm. C. Hood and Tjalling C. Koopmans, Yale University Press, 1953.

Koutsoyiannis, A., *Theory of Econometrics*, MacMillan, London,1973.

Van Leeuwen, I., Paelinck, J. and Wagenaar, Sj., *Towards Estimation of the Static Interregional Attraction Model for the Netherlands: an application of spatial econometrics, Foundations of Empirical Economic Research,* NEI, Rotterdam, 1976.

Marschak, Jacob, 'Economic Measurements for Policy and Prediction', in *Studies in Econometric Method*, ed. Wm. C. Hood and Tjalling C. Koopmans, Yale University Press, 1953.

Paelinck, Jean H.P., 'Modèles de Politique Economique Multirégionale Basés sur l'Analyse d'Attraction', *l'Actualité Economique*, octobre-décembre, 1973.

Van Wickeren, Alfred C., *Interindustry Relations: Some Attraction Models*, N.V.v/h Fa. M.J. van der Loeff, Enschede, 1971.

Wonnacott, Ronald J. and Wonnacott, Thomas H., *Econometrics*, John Wiley and Sons, New York, 1970.

7. Some considerations on labour market models (SPAMOL)

Introduction

With the treatment of a labour market model we simultaneously should attack the problem of commuting and the problems around migration. This means that a labour market model is in fact a model explaining the location of the supply of labour and the commuting performed by the workers.

It is implied in the foregoing statement that, with the treatment of even the most simple labour market model a beginning should be made with the integration of the models treated so far. In chapter 5 we studied models that relate to the amenities (including the quality of the environment). In chapter 6 where the attraction model was studied, the distribution of industrial and service activities over regions was determined. Now we will have to bring labour demand as resulting from the distribution of industry and labour supply as partly determined by the demand for labour and partly by the level of the environment and amenities together in order to arrive at an acceptable explanation of the distribution of labour supply and commuting patterns.

Labour supply

We will start from the assumption that all labour is supplied from the residences of the workers so that, assuming equal participation rates, the spatial distribution of the population is identical to the distribution of the labour supply. We further assume that population is attracted to a region as a result of three factors, viz. labour demand, quality of the environment and level of amenities.

Labour demand is defined as the availability of jobs in the own region and other regions, properly weighted with generalised transportation costs of commuting. Generalised transportation costs are defined as a weighted sum of money costs, time costs and comfort.

We may consequently use as a first definition for this labour demand potential

$$\pi_i^{D'} = \sum_j d_j e^{-ac_{ij}} \qquad (7.1)$$

in which d_j is labour demand in region j, c_{ij} is generalised transportation costs from region i to region j and a is the reciprocal of commuting mobility. If $a = 0$ distance does not play any role in commuting (mobility is infinitely high) and if $a = \infty$, $\pi_i^{D'} = d_i$ which simply means that only the jobs offered in the own region are of importance for the labour attractiveness of that region. It is assumed that the regions are measured in such a way that generalised transportation costs are zero for all practical purposes within the region.

However, since the function should have the character of a distribution function, the sum of all potential demands in all regions should equal actual demand. If we introduce this, we should rewrite (7.1) as

$$\pi_i^D = \sum_j \phi_j^D d_j e^{-ac_{ij}} \tag{7.2}$$

Writing

$$\sum_i \pi_i^D = d_j = \sum_i \sum_j \phi_j^D d_j e^{-ac_{ij}}$$

we find for

$$\phi_j^{-1} = \sum_i e^{-ac_{ij}} \tag{7.3}$$

so that

$$\pi_i^D = \sum_j d_j \frac{e^{-ad_{ij}}}{\sum_i e^{-ac_{ij}}} \tag{7.4}$$

or, in matrix notation

$$\pi^{\mathbf{D}} = \mathbf{A}\hat{\phi}^{\mathbf{D}}\mathbf{d} \tag{7.5}$$

in which $\hat{\phi}^{\mathbf{D}}$ is a diagonal matrix with $(\sum_i e^{-ac_{ij}})^{-1}$ as a typical element and **A** is the distance factor matrix with $a_{ij} = e^{-ac_{ij}}$ as a typical element.

In the most simple case when the supply of labour is determined by exogenous labour demand the following equation would hold

$$\mathbf{s} = \mathbf{A}\hat{\phi}^{\mathbf{D}}\mathbf{d} \qquad (7.6)\ [1]$$

The excess supply of labour for each region may in this case be written as

$$s - d = A \hat{\phi} d - d = [A \hat{\phi}^D - I] \, d \qquad (7.7)$$

This expression becomes zero for $A \hat{\phi}^D = I$ which is the case if $A = I$, in other words, if the resistance of distance is such that everybody works in his own region.

A maximum is reached if $A = I$ (I is the full unit matrix) which is the case if commuting mobility is infinitely large ($a = 0$). Population is in this case equally distributed over all regions, completely independent of labour demand.

Numbers of workers by origin and destination are presented in the matrix S

$$S = A \hat{\phi}^D \hat{d} \qquad (7.8)$$

The number of intraregional commuters is represented by the diagonal of **S** so that the matrix of the interregional commuters can be written as

$$C = S - \hat{S} = [A - I] \hat{\phi}^D \hat{D} \qquad (7.9)$$

in which $\hat{D}$ is the diagonal matrix of total labour demand and $\hat{S}$ of total labour supply.

Of course, there is no commuting if $A = I$. If distance does not play a role, $a = 0$ and consequently $A = I$. In that case total commuting equals

$$i' c_i = \frac{n-1}{n} d^{TOT} \qquad (7.10)$$

where d^{TOT} stands for total labour demand.

It may be reminded here that in this case working population is equally distributed over all regions.

We will now introduce the assumption that the supply of labour is not only attracted by job opportunities but as well by two other factors, viz., the supply of amenities and the environmental qualities of the region. We will represent these factors by π^a and π^e respectively.

A linear equation for the supply of labour then would become

$$s = a_1 \pi^a + a_2 \pi^e + \lambda \pi^D \qquad (7.11)$$

in which $a_1 + a_2 + \lambda = 1$ (7.12)

and the potentials are defined in a similar way as π^D.

In (7.11) a_1 represents the relative weight of the amenities, a_2 that of the environment and λ that of labour demand. (7.11) may be written as

$$s = a_1 \pi^a + a_2 \pi^e + \lambda A \hat{\phi}^D d \qquad (7.13)$$

For the time being we will replace $a_1 \pi^a + a_2 \pi^e$ by $(1 - \lambda) \pi^r$ in which π^r is defined as

$$\pi^r = \frac{a_1}{1-\lambda} \pi^a + \frac{a_2}{1-\lambda} \pi^e .$$

(7.13) then becomes

$$s = (1 - \lambda) \pi^r + \lambda A \hat{\phi}^D d \qquad (7.14)$$

Now obviously λ is not independent of a. If $a = 0$ this means that commuting distances do not play any role. If this holds true, obviously job opportunities have no importance whatsoever for the choice of the residence. If $a = \infty$, at the other hand, this must imply that job opportunities get complete priority.

This implies that we may write as a reasonable approach

$$1 - \lambda = e^{-aa} \qquad (7.15)$$

so that

$$s = e^{-aa} \pi^r + (1 - e^{-aa}) A \hat{\phi}^D d \qquad (7.16)$$

If $a = 0$ (7.1) reduces to

$$s = \pi^r \qquad (7.17)$$

In this case commuting distances are irrelevant and the choice of the residence is completely determined by non-labour market factors.

If $a = \infty$ (7.16) changes into

$$s = d \qquad (7.18)$$

Labour market factors are decisive for the location of workers.

It is here that a well known problem presents itself. If commuting behaviour is, as before, represented by the functions used earlier, total employment available to each region is expressed by

$$e = A \hat{\phi}^D d \qquad (7.19)$$

The difference between demand and supply, which is either open demand or unemployment is then given by

$$s - e = e^{-a\alpha} [\pi^r - A \hat{\phi}^D d] \tag{7.20}$$

for which, of course, again holds

$$i' s - i' e = 0 \tag{7.21}$$

The essential element in the foregoing argument is that workers express their preference for a residence in the area of their choice but do not accept the fact that this requires a different commuting pattern than the one assumed to hold if full employment is to be maintained. This means that, although the potentials are decisive for the location, commuting takes place on another basis than the judgement of job opportunities. Only if the definition of the potential itself is changed in such a way that the propensity to commute increases if the residential attraction of the own region is high, full employment can be maintained.

We then could write

$$\pi = \hat{\pi}^r A \hat{\phi}^r d \tag{7.22}$$

in which a typical element of

$$\hat{\phi}r = \frac{1}{\sum_i \pi_i^r e^{-ac_{ij}}} \tag{7.23}$$

so that the potential now is defined as [2]

$$\pi_i = \sum_j d_j \frac{\pi_i^r e^{-ac_{ij}}}{\sum_i \pi_i^r e^{-ac_{ij}}} \tag{7.24}$$

which means that the willingness to commute is a positive function of the residential qualities of the own region.

(7.16) now may be rewritten as

$$s = \hat{\pi}^r A \hat{\phi}^r d \tag{7.25}$$

In case distance prevents all commuting ($a = \infty$) we find $s = d$ (7.26)

In this case there is equilibrium between supply and demand in each and every region.

In case distance does not play any role at all

$$s = \pi^r \tag{7.27}$$

Location takes place on the basis of the residential quality only.

Labour demand

The foregoing analysis had as a basic assumption that the supply of labour adjusted itself completely to the demand for labour. The mechanism in this process was commuting which was such that the demand for labour was always satisfied by supply from all regions. It is obvious that there is in this case no particular need for a separate treatment of labour demand. The assumptions made imply that where ever the industries are located, they always will find labour in sufficient numbers available. The complete model thus may be written as

$$s = \hat{\pi}^r \ A \ \hat{\phi}^r \ d \tag{7.25}$$

$$d = d_{exo} \tag{7.28}$$

A refinement we could introduce in this model in order to make it somewhat more realistic is to assume that certain activities are oriented to the local market. In writing

$$d = d_{exo} + \beta \ \hat{y} p \tag{7.29}$$

indicating that some industries are exogenous and others attracted by local demand (here simply indicated as a fraction β of total local income yp in which p is population, and assuming

$$s = \gamma \ p \tag{7.30}$$

in which γ is the activity rate, we find

$$d = d_{exo} + \frac{\beta}{\gamma} \ \hat{y} \ s \tag{7.31}$$

so that (7.25) changes into

$$s = [I - \frac{\beta}{\gamma} \ \hat{y}]^{-1} \ \hat{\pi}^r \ A \ \hat{\phi}^r \ d_{exo} \tag{7.32}$$

In this model it would not make any sense to speak of labour-oriented industries since labour supply is available anywhere. However, commuting means a sacrifice and might follow from all sorts of decisions concerning the family, the environment for the children etc. which are taken in favour of others but for which the workers in the family will have to bear the consequences. Commuting may therefore, in spite of the formal equilibrium on the labour market be considered as a discrepancy which should be restricted as much as possible.

However, we should realise that commuting always will take place, however rational demand for labour is distributed over regions.

If all regions are equally attractive for living, and all regions have the same labour demand, the elements in the commuters matrix all reach their minimum value. Total commuting costs made by commuters in this case equals

$$c = i' \, [S^{MIN} \circledS \, C] \, i \tag{7.33}$$

in which c_{ij} is a typical element of **C**.

Since

$$S^{MIN} = \frac{d^{TOT}}{n} \; A \, \hat{\phi}^{D} \tag{7.34}$$

in which d^{TOT} is total demand for labour in all regions, (7.33) may be written as

$$c^{MIN} = \frac{d^{TOT}}{n} \cdot i' \, [A \; \phi^{D} \circledS \; C] \; i \tag{7.35}$$

The actual total distance covered equals

$$c = i' \; [S \circledS \; C] \; i = i' \; [\hat{\pi}^{R} \; A \; \hat{\phi}^{R} \, \hat{D} \circledS \; C] \; i \tag{7.36}$$

so that

$$c^{X} = c - c^{MIN} = i' \, [\, [\hat{\pi}^{R} \, A \, \hat{\phi}^{R} \, \hat{D} - \frac{D^{T}}{n} \cdot A \, \hat{\phi}^{D}] \circledS \; C] \; i \tag{7.37}$$

represents excess-commuting measured in terms of total excess distance covered.

It appears that c^{X} decreases as both labour demand and amenities are more equally distributed and as *a* increases, or, which is a more realistic case, if all transportation costs increase proportionally. We will come to the application of this model later.

Generalised transportation costs; a feedback

So far, generalised transportation costs represented by c_{ij} were assumed to be exogenously given. This does in fact not seem very realistic. These transportation costs include the value of time and comfort implied in commuting which in general will be dependent on the volume of the transportation flow. In a general model this assumption might hold to a certain extent, as long as the investments in roads adjust to the capacity of the roads in such a way that transportation costs on each part of the network remain constant. In cities and more particularly in city centres, however, this does not apply since possibilities for road construction are lacking and it may in some cases even be a firm decision not to give way to traffic flows but to increase transportation costs deliberately in order to decrease traffic flows or to influence the modal split in favour of mass transportation.

It appears that in this case the model becomes rather complicated. This may be shown as follows:

A reasonable function representing the relation between transportation costs on a given road and the volume of traffic is

$$c_{ij} = \bar{c}_{ij} \, e^{\delta s_{ij}} \tag{7.38}$$

Combining this function with the basic function we obtain

$$s_{ij} = \frac{d_j e^{-a\bar{c}_{ij} e^{\delta s_{ij}}}}{\sum_i e^{-a\bar{c}_{ij} e^{\delta s_{ij}}}} \tag{7.39}$$

which is not particularly easy to handle even in this simple case in which it is assumed that each ij-commuting takes place over a separate ij-road connection. Since we want to concentrate our attention later on city traffic in which a very considerable part of commuter traffic is directed towards the centre from different other parts of the city, this assumption is not too unrealistic for our problem. However, it appears that even in this case the solution to the system of equations presented in (7.39) is not an easy one. For the solution iterative methods will have to be applied.

Urban location

In order to look somewhat more closely at the situation in urban areas we will take a simple case in which one region is the city centre and another one the rest of the city. In such a situation the condition is fulfilled that each commuting flow has its own road-link. The equations for this situation may be written as

$$s_1 = \frac{\pi_1^r}{\pi_1^r + a_{21}\pi_2^r} d_1 + \frac{a_{12}\pi_1^r}{a_{12}\pi_1^r + \pi_2^r} d_2 \tag{7.40}$$

$$s_2 = \frac{a_{21}\pi_2^r}{\pi_1^r + a_{21}\pi_2^r} d_1 + \frac{\pi_2^r}{a_{12}\pi_1^r + \pi_2^r} d_2 \tag{7.41}$$

We assume that there is congestion on the roads to the city in the morning and in the evening from the city. In the reverse direction there is no congestion. This means that d_{21} is high compared to d_{12} or in terms of the equations written down a_{21} is small compared to a_{12}.

A further decrease in a_{21} may be calculated as

$$\frac{-\partial s_1}{\partial a_{21}} = \frac{-\pi_1^r \pi_2^r}{(\pi_1^r + a_{21}\pi_2^r)^2} d_1 \tag{7.42}$$

$$\frac{-\partial s_2}{\partial a_{21}} = \frac{\pi_1^r \pi_2^r}{(\pi_1^r + a_{21}\pi_2^r)^2} d_1 \tag{7.43}$$

which means that location in the city centre is encouraged and in other areas is discouraged by a further increase in transportation costs towards the centre. In the preceding paragraph we indicated already that there is in fact an equilibrium situation since transportation costs themselves are a function of the volume of traffic. However, what we have in mind here is an autonomous increase in transportation costs through government measures like road pricing, or a rise in parking rates in the city centre.

A second measure that could be taken is the improvement in living conditions in the centre. This can be reached by increasing the level of π_1^r. The results are:

$$\frac{\partial s_1}{\partial \pi_1^r} = \frac{a_{21}\pi_2^r}{(\pi_1^r + a_{21}\pi_2^r)^2} d_1 + \frac{a_{12}\pi_2^r}{(a_{12}\pi_1^r + \pi_2^r)^2} d_2 \qquad (7.44)$$

$$\frac{\partial s_2}{\partial \pi_1^r} = \frac{-a_{21}\pi_2^r}{(\pi_1^r + a_{21}\pi_2^r)^2} d_1 - \frac{a_{12}\pi_2^r}{(a_{12}\pi_1^r + \pi_2^r)^2} d_2 \qquad (7.45)$$

It appears that the influence of an improvement of weighted living conditions in the centre is stronger, the higher the value of π_2^r, in other words, the better the conditions elsewhere are.

The third possible measure is the relocation of employment to the outskirts of the city. We rewrite (7.40) and (7.41)

$$s_1 = -\left\{\frac{\pi_1^r}{\pi_1^r + a_{21}\pi_2^r} - \frac{a_{12}\pi_1^r}{a_{12}\pi_1^r + \pi_2^r}\right\} d_1 + \frac{\pi_1^r}{\pi_1^r + a_{21}\pi_2^r} d \qquad (7.46)$$

$$s_2 = -\left\{\frac{a_{21}\pi_2^r}{\pi_1^r + a_{21}\pi_2^r} - \frac{\pi_2^r}{a_{12}\pi_1^r + \pi_2^r}\right\} d_1 + \frac{a_{21}\pi_2^r}{\pi_1^r + a_{21}\pi_2^r} d \qquad (7.47)$$

The coefficients of d_1 represent the influence of a change in d_1. They may be rewritten as

$$\frac{\partial s_1}{\partial d_2} = -\frac{(1 - a_{12}a_{21})\,\pi_1^r}{(\pi_1^r + a_{21}\pi_2^r)\,(a_{12}\pi_1^r + \pi_2^r)} \qquad (7.48)$$

$$\frac{\partial s_2}{\partial d_2} = \frac{(1 - a_{12}a_{21})\,\pi_1^r}{(\pi_1^r + a_{21}\pi_2^r)\,(a_{12}\pi_1^r + \pi_2^r)} \qquad (7.49)$$

which shows a considerable negative effect on the location of workers in the city and an equally large positive effect on location elsewhere.

Spatial discrepancies

In the foregoing sections we treated two basically different approaches to commuting behaviour. The first was that this pattern is independent of the spatial distribution of employment. The result was unemployment on the one hand and open demand on the other. The second approach assumed that the willingness to commute was a function of the preferences of the consumer for certain residential qualities of the different regions, in such a way that full employment was maintained. The stronger the preferences to live in a given area, the stronger also the willingness to commute.

Actual behaviour will lie somewhere in between these two extreme cases. A complete independence of commuting from the location of job opportunities is as unlikely as a complete adjustment to them. Particularly in peripheral areas of countries, unemployment is as normal a phenomenon as open demand is in congested areas (under 'normal' economic conditions). Preferences for living in some region either as a result of excellent living conditions or reluctance to leave the region might be such that even if there is a certain willingness to commute over large distances, unemployment and open demand will be the inevitable consequence.

However, even if unemployment and open demand demonstrate themselves as the most obvious consequences of the differences in spatial structure of demand and supply for labour, they are by no means the only consequences. Those who work will at least partly have to make long daily trips to reach their work, while others will be employed in professions for which they are not trained. Although both groups are actually employed this does not mean that their situation is ideal. Far from that.

The conclusion should be that it would, from an analytical point of view, be unwise to speak of employed persons and unemployed persons only. It rather seems appropriate to express the situation in a tension-coefficient indicating the degree to which the actual situation deviates from the desired situations. Indicating the latter by $\tau_o^c = e^{-ac_{ii}} = 1$, the tension coefficient for an individual becomes

$$\tau_{ij}^c = 1 - e^{-ac_{ij}} \qquad (7.50)$$

and for the labour market as a whole

$$\tau^c = \sum_i \sum_j s_{ij}^c \,(1 - e^{ac_{ij}}) \qquad (7.51)$$

$$0 \leqslant \tau^c \leqslant 1$$

or in matrix notation

$$\tau^c = i' [S^c - (S^c \circledS A^c)] i \qquad (7.52)$$

in which $\circledS$ is the Schur-product and a^c_{ij} is the typical element of the distance matrix A ($a^c_{ij} = e^{-ac_{ij}}$).

τ^c becomes equal to zero if all c_{ij} become zero ($A^c = I$) or if $a = 0$. Both cases represent an ideal situation in which either all distances between jobs and workers become zero (not a very relevant case in regional science!) or mobility is infinitely large.

Professional discrepancies

Just as one might consider the distance between regions, one could consider the 'distances' between professions. In order to analyse this problem we might in principle use a model similar to that presented in the preceding section.

Let

$$s_1 = s_{11} - s_{12} + s_{21} \qquad (7.53)$$

$$s_2 = s_{22} - s_{21} + s_{12} \qquad (7.54)$$

be the potential supply of workers in profession 1, consisting of those trained for profession 1 and working in profession 1 (s_{11}), those trained for profession 1 but working in profession 2 (s_{12}) and the reverse case s_{21}, being the number of workers skilled in profession 2 but working in profession 1.

We assume

$$s_{kl} = \frac{d_1 e^{-\beta\delta_{kl}}}{\sum_j d_1 e^{-\beta\delta_{kl}}} s_i \qquad (7.55)$$

represents the probability that someone trained for job 1 will work in job 2. This probability is a function of the demand for jobs 1 in relation to all other demands, as well as of all professional 'distances'.

It may easily be shown that maximum unemployment and shortage of certain skills results, if all $\delta_{kl} = \infty$ or $\beta = \infty$, and full employment if all

$\delta_{kl} = 0$, or $\beta = 0$. The latter case represents the one in which professional mobility is infinitely large, the first case if each and every worker refuses to accept a job out of his precisely defined own job.

Again, if we define the optimum situation as one in which everybody works in his own job, we may construct a tension coefficient analogous to the first one. We then write:

$$\tau^p_{kl} = 1 - e^{-\beta\delta_{kl}} \quad (7.56)$$

and thus for the labour market as a whole:

$$\tau^p = \sum_k \sum_l s_{kl} (1 - e^{-\beta\delta_{kl}}) \qquad 0 \leqslant \tau^p \leqslant 1 \quad (7.57)$$

or in matrix notation

$$\tau^p = i' [S^p - S^p \circledS A^p] \, i \quad (7.58)$$

A general tension coefficient

A general coefficient may be derived from the two foregoing sections, combining both commuting distances and professional distances.

We define for the individual

$$\tau^{kl}_{ij} = 1 - e^{-(\alpha c_{ij} + \beta\delta_{kl})} \quad (7.59)$$

as the general tension coefficient. For the labour market as a whole this coefficient becomes

$$\tau = \sum_i \sum_j \sum_k \sum_l s^{kl}_{ij} \left\{ 1 - e^{-\alpha c_{ij} - \beta\delta_{kl})} \right\} \quad (7.60)$$

which becomes equal to unity if and only if all αc_{ij} and all $\beta\delta_{ij}$ become zero.

Labour market and regional policy

The foregoing analysis shows that discrepancies or tensions on the labour market exist to a far larger degree than indicated by unemployment figures or open demand figures. However, one might argue that this is simply the result of the fact that a general equilibrium is incompatible to a certain degree, with partial equilibria for each of the elements. If, one could argue, the commuting distance for an individual becomes large because he wants to live in a certain region which he prefers because of environmental reasons or the supply of amenities there and simultaneously he insists to work in a very well-defined job which is only available in a region at considerable distance from the one which he has chosen as his residence, then a large commuting distance is the inevitable result. The sacrifices to be made to bridge this distance each workday, however, are in this case over-ruled by the advantages in living. The choice made is perfectly rational and follows from the fact that working- and living conditions together determine his welfare. Improving his working conditions by moving closer to his job would decrease the discrepancy on his private labour market but at the same time decrease his well being. Therefore, the actual situation is the best and should not be altered.

Seen in this way, one might say that, for example, discrepancies on the labour market, resulting from the fact that demand-factors and supply-factors do not coincide in space, are a quite natural phenomenon, a result of the search for optimal living conditions of the individuals who prefer a certain environment for their residence so much more than another, that they gladly accept the consequences in the form of a certain commuting and/or unemployment and the advantages of their residence over those of another location where well being would be less. Seen from this angle the situation is quite acceptable.

However, one might argue as well that a different situation, where some regions are made more attractive by government measures for protecting the environment, better housing conditions, better and more amenities etc. and other regions are made more attractive for industrial activities by improvements of infrastructure, or government grants on investments, a better balance between demand and supply per region could be obtained than if everything is left to its natural course. If this view is acceptable, then it does seem useful to consider commuting as a discrepancy on the labour market and consequently, a situation with smaller commuting distances as a better situation than one with larger commuting distances.

Moreover the situation could be improved by a government infra-structure policy which could lead to a decrease in general transportation costs (c_{ij}), mitigating in this way directly, the disadvantages of commuting.

For the professional side of the problem similar arguments hold. Government influence could be used to encourage industry to use certain skills in over-supply, to train young people more in jobs that will be in demand in the future, to improve professional mobility by retraining courses and improved systems of training, enabling workers to shift more easily from one job to another.

All these measures could be taken to lessen the tensions on the labour market and implicitly contribute to higher total employment and smaller total open demand.

Notes

[1] This expression may also be written $\mathbf{s} = \mathbf{A}\,[\widehat{\mathbf{A}'\,\mathbf{i}}]^{-1}\,\mathrm{d}$, the sign $\wedge$ meaning diagonalisation.

[2] Which is a special case of the more general function

$$\pi_i = \sum_j d_j \frac{(\pi_i^r)^\beta \, e^{-ac_{ij}}}{\sum_i (\pi_i^r)^\beta \, e^{-ac_{ij}}}$$

References

Heijke, J.A.M., Klaassen, L.H. and Offereins, C.J., *Naar een arbeidsmarktmodel*, Tjeenk Willink, Groningen, 1974.

Klaassen, L.H. and Drewe, P., *Migration Policy in Europe*, Saxon House/ Lexington Books, Westmead, Farnborough, 1973.

Klaassen, L.H. and Heijke, J.A.M., *Some Indicators of Regional Labour-market Equilibrium*, Series Foundations of Empirical Economic Research, NEI, Rotterdam, 1975/2.

Kornai, J., *Anti-equilibrium; on Economic Systems Theory and the Tasks of Research*, North-Holland Publishing Company, Amsterdam/London, 1971.

8. Some Dutch experiences with the use of integrated spatial models [1]

Introduction [2]

After having considered a series of partial models as well as a framework for an integrated model it seems worthwhile to put the question whether or not it is possible to put the partial models together into one integrated model and, more specifically, to try to find an answer to the question whether such a general integrated model would be an effective instrument in the hands of the government, in its efforts to effectuate a general societal policy. This final chapter is dedicated to this problem.

When Tinbergen wrote his excellent book on Economic Policy, [3] *Principles and Design*, a new kind of economic policy was born. By means of a system of simultaneous equations the economic system as a whole was described in quantitative terms, and instrument variables forming an important part of the total set of variables, were supposed to be used by governments to steer the process of economic development and enabled them to attain, at least under certain conditions, the desired level of the objectives or goal-variables.

The most fascinating features of this system were its simplicity and clarity. It seemed that the only thing left to be done was to build a model that would adequately describe the interrelations between the relevant variables. Once this problem was settled, there would be no major difficulty to stand in the way of efficient and effective economic government policy.

This chapter tries to analyse how society has developed since this system of economic policy was introduced and, more specifically, to find out if it is still acceptable in the world of today. (It may be said here that Tinbergen himself has repeatedly, either implicitly or explicitly, expressed doubt regarding the applicability of his original system). Furthermore, an attempt will be made to establish in some detail what, if any, 'the' model in regional science would look like, should the 'classical' economic policy model no longer appear relevant.

The principles of the theory of economic policy, reconsidered

The theory of economic policy was based on four hypotheses:

1 There is a relatively simple system of equations that describes adequately the main features of economic development.

2 There is a definite and limited set of goal variables.

3 The government is free to use the available instruments and to introduce new instruments.

4 Subject to certain conditions, the desired values of the goal variables can be attained by the proper use of instrument variables.

It seems useful to study all four hypotheses somewhat closer, so as to find out to what extent they offer a realistic approach to regional science. That is the more important as the model resulting from the four hypotheses is still widely used, also in regional studies.

The system of equations

The difference between the models used in general economics and as yet hypothetical models we are favouring nowadays, is essentially two-fold.

First, a new dimension has been added to economic analysis by the introduction of elements of space or distance. Now the addition of a new dimension to a science, an element that used not to be there, by definition should render that science more general and also, for that matter, more realistic. Yet, in many schools regional economics is still treated as a specialisation of 'general' economics, as a specific daughter of the motherscience. Why that should be so is difficult to say: perhaps one reason might be the lack of data on the regional level, which forces regional economists to limit their analyses to only parts of the field covered by general economics, and to leave out, for example, monetary questions; even studies of investment and consumption behaviour are often out of their reach. So, they are apt to concentrate on industrial structures, employment analysis, migration of households, and similar studies. Still, with data on other variables becoming increasingly available, the scope of the analyses tends to widen. Be that as it may, the principle stands that regional economics is a more general discipline than general economics, and therefore requires considerably larger models than have ever before been used in economics.

The second difference between the modern and the conventional approach to economics lies in our present conviction that we shall never be able to explain developments in the real world by a mono-disciplinary economic approach, and that other disciplines will have to be incorporated in our analyses if we want to construct realistic models.

Of other disciplines there are many, of course, but to achieve a more or less general model, environmental issues, transportation issues, and social issues should at the least be taken into account.

That environmental issues should be incorporated in our analysis is obvious nowadays. The deterioration of environmental qualities has become an important factor not only in investment decisions, but also in location decisions of private households. Besides the more or less objectively measurable deterioration of the environment, the increased sensitivity of people vis à vis their environment must very likely be held responsible for the great weight nowadays attached to environmental problems.

The need to study transportation issues follows logically from the introduction of spatial factors in the analysis; moreover, particularly in larger cities, traffic is one of the fundamental factors in the development of urban structures, and one that in many cases has a decidedly unfavourable influence.

Social issues, finally, have gained in importance as thinking on distributional problems progressed. Distribution of income, of power and knowledge in many countries has become a major issue of general government policy, and thus its incorporation in a general model has become essential.

Needless to say that both the introduction of the spatial elements and the enlisting of new disciplines have dramatically changed the size and scope of the original models. Without exaggeration a factor hundred can be considered a very modest estimate of the increase in the number of variables.

The set of goal variables

In a classical economic model the set of goal variables was limited to four or five. They accounted for a sizeable growth of the national product, equilibrium in the balance of payments, full employment, and a reasonably stable price level, all of these measured on the national level.

It follows logically from the foregoing that in a regional approach to the problems all the goals formerly set on the national level should now be set on the regional level. Consequently the number of goal variables is to be multiplied by the number of regions considered in a country, of course with proper adjustment of the balance-of-payment concept. Furthermore, the introduction of the social issues, with the attendant distribution aspects, leads not only to disaggregation of the variables into subvariables for different groups of the population, but as a matter of fact also to a disaggregated set of goal variables.

If we consider just three income groups, five social groups, and ten regions, we should be faced with 150 times as many goal variables as had to be set out on the national level for the population as a whole, at least as far as these goals are also relevant on the lower levels of disaggregation.

In addition to the huge problem thus created, another one, no less important and certainly no less difficult, has arisen. Partly in response to actual developments, partly as a return to the original basic ideas of economic science, the concept of individual welfare was replaced by the concept of well-being, a general expression for the weighted sum of all things pleasant and unpleasant experienced by the population of a country. No longer do we measure the results of all developments in one single figure such as the level of personal income, or consumption; nowadays we express them in a complex figure in which all regions, all population groups, and all elements of their well-being are accounted for. This complex figure, this super goal, is what we want to optimise, with the help of our huge, and still vigorously growing, models. And sometimes we should do so with the help of the instrument variables in our models.

The set of instruments and the freedom to use them

Looking closely at the set of instruments at the disposal of the governments, we find the idea of describing the whole society with the help of one enormous model increasingly doubtful. After what has been said earlier it will be obvious that the more goals are set, the more instruments will be required to steer development in the desired direction. Now if the government were free to use every possible instrument, and if the total general model, properly tested, were available, *and* if the government had a clear idea of what well-being is in quantitative terms, then the government could hope to help increase well-being. Alas, the situation is not like that. We don't have the model and we probably never will have it, and the government does not have an idea, most certainly not in quantitative terms, what well-being is, but above all, the government is not free and is even bound to become less and less free to use potential and traditional instruments. And in many cases it is the governments themselves who want it to be that way. The redistribution of well-being includes the redistribution of power and knowledge, precisely the two factors that make for extra-governmental influence on the course of events in our society.

Before going further into this, another important fact is to be registered. The introduction of the element of space or distance into our science implies in practice the introduction of regions as elements in our models and consequently the introduction of the influence of regional and local governments. Although there are, of course, differences from country to country in the relative power of those regional and local governments, there is always a certain degree of autonomy, implying the power to use instruments on the regional and local levels. It is unlikely that regional and local authorities will always use the policy instruments at their disposal in the way the central government would, for their super goal is regional or local well-being, definitely not national well-being. Their views are more limited than that of the central government, as, of course, they are supposed to be. No citizen would like it very much if his local government sent sizeable amounts of money to other regions in order to contribute to national well-being, thus leaving local problems unsolved. In a democratic system such a government is likely to be replaced very quickly with a government with a somewhat narrower view, taking more care of the problems at home. In addition to these governments of lower order, international governments like, for example, the EC are gradually gaining weight and limiting the power of national governments. Subsidies on production, devaluations or revaluations of the currency, are no longer instruments solely to be used by national governments. Though, of course, not wholly powerless, the national government has at least to consult the relevant international organisations before taking steps in one of these fields.

It would be erroneous to think, however, that only other governments limit the power of the national government. On the market of power, national governments find many competitors, gradually growing in importance and gradually making it more and more difficult for the government to do what they think right and proper in the national interest. These are the labour unions, the employers' unions, the environmentalists, the car owners, the house owners, the women's organisations, the students, the senior citizens and all other groups having become active participators in decision making. Even regional groups, such as the population of underprivileged regions or of 19th century town districts, are raising their voices in defence of their own position and rights. They often enter into a public discussion with the government, either directly or indirectly through their representatives in parliament, often successfully interfering with government plans relating to the interests they represent.

That is a development which, particularly in the last decade, has considerably modified decision making processes in our countries, often restricting very severely the influence the government can exert on the course of events in society. The ideal picture of the wise and fatherly governors who know what is good for us has almost completely vanished. And even if the government knew what is good for us, they are often unable to bring it about, because the very instruments are often denied them. And even if they could use the proper instruments it is doubtful that they would know how they were going to work out. The time of simple steering mechanisms is over.

The sub-model approach

The considerations presented in the foregoing reflect the authors' doubts concerning the possibility of constructing a general disaggregated model that also could be used as a basis for government policy. If these doubts are justified, we are faced with two important questions, viz.

1. If general models in which all relevant variables play their proper roles are so complex that our present knowledge is inadequate to construct them successfully, how could we then construct meaningful sub-models? (that, at least to a certain extent, could be used.)
2. If we were able to construct meaningful sub-models, would they be relevant also for government policy? In other words, is the best way to break down the general model also the best way to structure government policy?

The first question touches in fact the problem how to break down the general model into sub-models without violating too much the main object of the general model, viz. registration of all interactions between all variables. When we cut out a piece of the large model, we implicitly cut interaction flows (one way as well as two way flows) between the variables in the sub-model and those in the remaining part of the general model. The first and general principle should be, then, to cut the sub-models out of the general model in such a way that a minimum of interaction flows are interrupted.

The second principle is based upon the assumption that the intensity of interaction decreases with physical distance. This is supposed to hold not only for consumer activities such as recreation, shopping, or more general, the use of social infrastructural elements, but also and just as much for deliveries by industrial firms and inter-industrial relations. The larger therefore the regions chosen as a basis for the analysis, the smaller the size of the interaction flows will be in relation to the total size of intraregional activities.

The final principle derives from the fact that the distance-decay-function is different for each activity. For that reason it is essentially possible to rank each activity according to the influence distance exerts on the demand for its products and/or the supply of its inputs. Obviously this influence is smaller with most services, particularly those for which the consumer has to move in person to the place of production, than with industrial products, which can be transported over larger distances at relatively low cost. In a ranking, we shall find most industrial activities in the bracket of low distance elasticities and most service activities in the high distance-elasticity bracket. High distance-elasticity activities may be called low-order activities, and activities characterised by low distance elasticities may be denoted as high-order activities.

If we start with relatively large regions we shall find that the flow of goods and services passing the boundaries consists mainly of flows resulting from high-order activities, and that they are characterised by a high degree of autarky as far as lower-order activities are concerned. Low-order activities are, in fact, irrelevant when we study development of and the interrelations between large regions.

It seems natural to use the results of studies of these larger regions (such as industrial location studies), as inputs in studies of medium sized regions, making them serve as quasi exogenous variables. [4]

The same process can be repeated in a study of smaller regions, the results obtained for the medium-sized regions serving in turn as quasi exogenous variables. In this way a hierarchy of models is created in which each activity, according to the size of its distance elasticity, is studied in a relevant regional context.

The foregoing implies that for a proper division into large, medium sized, and small regions, the distance elasticities of the different activities need to be studied. In practice, because of the structure of the available data, one is often forced to use existent administrative divisions. That is a disadvantage in as much as such a division might not correspond with a set of different-sized relevant regions, but an advantage in as much as such a division usually corresponds with administrative boundaries. Regular adjustments of these boundaries, particularly the municipal ones, prevent the differences between administrative and analytical regions from becoming prohibitive to a proper analysis.

If we assume further that a government on a given level (de facto the government of a region of a given size) is responsible for all activities of the corresponding or lower levels, which means in practice all activities for which the relevant region is equal to or smaller than the region for which this government is held responsible, we have implicitly answered the second question put forward in the beginning of this section. In this construction the regional government is not responsible for activities with large external effects, since these, by definition, fall under the jurisdiction of a higher-level government.

In the foregoing approach the general model was, in a logical way, broken down into at least three sub-models, each relating to regions of a given size and implicitly relating to corresponding groups of activities of different 'levels'.

Urban models

It would be a bit too optimistic to assume that with this disaggregation into three or at most four groups of models, the general model could be tackled easily. Since Lowry's efforts it has become clear that in particular studies on the level of local government (including that of urban regions) are far too complicated to be handled by simple techniques. [5] That is due to the fact that urban regions besides containing the full range of low-level activities, also serve as important centres of higher-order activities, thus mixing activities with low, medium, and high distance elasticities.

Unlike the larger regions, the physical structure, i.e. the way they are structured spatially, is an important element with urban regions. What with the high population density in these areas and the intensive interactions between the population groups, huge traffic flows are created by commuting, shopping, and recreation, to mention just the most important activities. The size of these traffic flows depend very much on the physical structure of the city, specifically on the spatial arrangement of workplaces, residential areas, and elements of social infrastructure.

Recent studies [6] are gradually making clear that the processes of urban development are so complex and the dynamic forces so strong that neither Eastern nor Western European governments have been able to control these processes properly. There are good reasons to believe that the situation is even worse, and that until quite recently, urban governments who were able to do some steering have done so in such a way as to stimulate an evolution passing from urbanisation via suburbanisation to disastrous desuburbanisation. Desperate efforts are now being made to restore the cities as the centres of our social and cultural life, but unfortunately only there where the process of desuburbanisation is already well under way. Urban policy is a fine illustration of how little we benefit from the experiences of others. Every city government considers its own problems unique and is only to a very limited extent prepared to learn from those who know where it all leads.

The reason for the extraordinary behaviour of city governments may again lie in the extreme complexity of the system for which they are responsible, the enormous difficulty of registering adequately what actually is going on, and the uncertainty about what instruments indeed contribute to the well-being of the different groups of the population. No need to say that another important factor is the limited financial means available to most urban governments.

The interesting question how regional science could contribute to a more acceptable (let us forget for the moment the word optimum) development of urban regions, is extremely difficult to answer, but a fruitful approach could perhaps be the intensive study of separate activities in sub-models integrating both the demand and the supply side. It is important to stress the necessity of treating both sides of the market of activities in such sub-models. Too many models of this type concentrate on just one side of the problem. Shopping models 'explain' consumer behaviour *given* the location of the shopping centre, commuting studies 'explain' commuting behaviour *given* the location of residential areas and concentrations of workplaces, recreation studies show how people react to the supply of given recreation areas located in given places etc.

We shall have to take one more step and try to explain as well why the shops are where they are, why people live in a given place and work in a given place. The problem of use and location are problems that will have to be solved simultaneously [7] in one single sub-model.

It is far from clear how all these sub-models could be integrated into a general urban model and, more specifically, how the government would influence the functioning of such a model in practice. However, as all studies indicated contribute in some way to an understanding of the traffic flows that so strongly govern the direction in which our cities are developing, more behaviouristic traffic models could perhaps be created with the help of integrated activity and allocation models, which should complement the conventional transportation models. These models could provide the urban governments with more insight and, hopefully, better instruments than the ones that are used today. Provided that urban governments know what they want or, rather, what in the long run the population of the city will want, these instruments could help to attain more sensible goals than have been set in the past.

It is important to note that in this view physical planning is not treated as a separate discipline to be applied separately from other disciplines. As we become able gradually to build a sensible general model that at least describes people's behaviour in the most important activities, including their resulting transportation behaviour, we shall implicitly come to understand the physical structure of the city. What remains to be done, of course, is incorporating the influence that the government could exert on the spatial development processes by means of zoning regulations, rehabilitation of city centres, creation of pedestrian zones, and limitation of uncontrolled suburbanisation.

Final remarks

In the foregoing we presented a limited view of the development of regional science and its significance for practical policy. We have seen how limited is the influence governments can exert on regional and urban development, a fact that should warn us against the lighthearted use many highly theoretical regional studies make of optimisation models, and make us aware of our duty to find out much more of how the real world functions before we presume to advise governments or to criticise them if they fail to heed our advise. Regional science ought not to claim having the answers to all regional problems.

We are all working away with a small teaspoon at the huge mountain formed by *the* regional problem. And *the* regional problem is learning to understand how our society functions.

Notes

[1] Presented as a Presidential Address to the Meeting of the Regional Science Association at Krakóv, August 1977 by Leo H. Klaassen.
[2] Compare W. Eizenga and L.H. Klaassen, 'Enkele gedachten over de grenzen van de macht van de overheid', ESB,1977, no.3104, p.464 ff.
[3] Amsterdam, 1956.
[4] See for example Willem Molle and Bas van Holst, *Factors of Location in Europe: A Progress Report*, FEER, 1976/15, Netherlands Economic Institute, Rotterdam. In this study Western Europe is divided into 100 regions, the location behaviour of 70 sectors is studied for 1950, 1960 and 1970. Inter-industrial relations play an important part in this study which has not yet reached its final stage.
[5] Compare M. de Langen, I.L. van Leeuwen and A.C.P. Verster, *The Influence of Accessibility on the Spatial Distribution of Activities in an Urban System: A Progress Report*, FEER 1977/4, Netherlands Economic Institute, Rotterdam.
[6] The CURB Project, a study on the costs of urban growth in 13 Eastern and Western European countries. Centre Européen de Coordination de Recherche et Coordination en Sciences Sociales, Vienna.

[7] An excellent presentation of the problem is given in A.C.P. Verster, *The interrelations between Residential Migration, Change of Jobs and Residence-work Accessibilities, A Research Model (Dutch).* Report Netherlands Economic Institute, Rotterdam, April 1977.

Index